CW01544225

STARCHILD

STARCHILD

MY LIFE UNDER THE NIGHT SKY

MAGGIE ADERIN

BBC BOOKS

UK | USA | Canada | Ireland | Australia
India | New Zealand | South Africa

BBC Books is part of the Penguin Random House group of companies
whose addresses can be found at global.penguinrandomhouse.com

Penguin Random House UK
One Embassy Gardens, 8 Viaduct Gardens,
London SW11 7BW

penguin.co.uk

First published by BBC Books in 2026

1

Copyright © Maggie Aderin 2026
The moral right of the author has been asserted.

Penguin Random House values and supports copyright. Copyright fuels creativity, encourages diverse voices, promotes freedom of expression and supports a vibrant culture. Thank you for purchasing an authorised edition of this book and for respecting intellectual property laws by not reproducing, scanning or distributing any part of it by any means without permission. You are supporting authors and enabling Penguin Random House to continue to publish books for everyone. No part of this book may be used or reproduced in any manner for the purpose of training artificial intelligence technologies or systems. In accordance with Article 4(3) of the DSM Directive 2019/790, Penguin Random House expressly reserves this work from the text and data mining exception.

Typeset by Francisca Monteiro
Printed and bound in Great Britain by Clays Ltd, Elcograf S.p.A.

The authorised representative in the EEA is Penguin Random House Ireland,
Morrison Chambers, 32 Nassau Street, Dublin D02 YH68.

A CIP catalogue record for this book is available from the British Library

ISBN 9781785949449

Penguin Random House is committed to a sustainable
future for our business, our readers and our planet.
This book is made from Forest Stewardship Council® certified paper.

For Lori,
my greatest adventure, my constant inspiration,
and the brightest star in every universe.

And for you, the reader,
thank you for walking beside me through this version of reality –
for sharing the wonder, the questions, and the light.

We are a way for the cosmos to know itself. – *Carl Sagan*

Contents

Thank you   ix

1. The Retrospectroscope   1
2. Boarding School Blues   41
3. Dunce Dunce Double D   75
4. The Chameleon   105
5. Doctor, Doctor   135
6. Getting to Work   159
7. Father, Brother, Teacher, Philosopher, Friend   175
8. Milestones   191
9. Starchild   235
10. The Many Maggies   256

Epilogue   303
Acknowledgements   309

Thank you

Every decision we make – to speak or stay silent, to turn left instead of right, to look up at the sky or down at our feet – each decision feels small, fleeting. Yet some physicists suggest that each choice might do more than shape our lives; it could shape reality itself.

In the many-worlds interpretation of quantum mechanics, every possible outcome of every event actually happens – each in its own universe. When you choose one path, another version of you chooses differently, and both futures unfold in parallel, forever divided by a single moment of possibility.

It's a theory born from science, but it feels like poetry: the idea that the universe is not one story, but an infinite library of them – each page turned by a decision, each life a thread in the cosmic tapestry of what *might have been*. I've often wondered how many versions of me exist out there in the great expanse – in the countless parallel worlds that physics dares us to imagine. Perhaps somewhere there's another Maggie who never looked up through a telescope, who took a different path entirely. Maybe she became a teacher, a traveller or simply someone who gazed at the stars and thought, *Not tonight*.

THANK YOU

The idea that every decision spins off a new reality – every 'yes', every 'no', every step forward or back – is both humbling and exhilarating at the same time. Somewhere, infinite Maggies are living out the infinite possibilities of all the choices I didn't make. But the one who sits here now, writing these words, is the version I know best.

And I am grateful for her.

Grateful for this universe, this improbable collection of moments that brought me here. Grateful for every person who crossed my path – family, friends, mentors and also the many strangers who offered kindness or challenge along the way. You've each shaped the trajectory of my world, bending my light in ways that have made it brighter.

If other versions of me exist, I wish them well. But I am happy to be *this* one – in *this* universe, at *this* point in time – surrounded by the people and the wonders that made my journey what it is.

Because when I look up at the night sky, full of stars and possibilities, I know that I wouldn't trade this version of reality for any other.

Chapter 1
The Retrospectroscope

For the last 50 or so years, pretty much my entire life, I believed I was born three weeks past my due date. In fact, I used this as the reason why I was always late for things. It's not my fault, I would think and sometimes say out loud, it's how I arrived in this world and it's how I have navigated through it. Eternally behind, shouting to everyone, 'Don't worry guys, I'll catch you up!'

It was only recently that my mother, sorting through old medical paperwork, did a double take and realised she had been wrong. She had muddled me up with one of my sisters; I had arrived bang on time. I was a bit miffed for a while. Not only did this put a different slant on the photograph of me being valiantly held aloft by my older sister, her skinny legs braced to hold the weight of a cannonball-heavy baby, but it also meant I had lost what I thought was the perfect excuse for tardiness. Most importantly, something I believed about myself for as long as I had existed turned out to be incorrect.

While this new information was hardly life-changing, it did make me think about my process of writing this memoir. I realised I needed to turn the spotlight on everything I thought to

CHAPTER 1

be true and investigate those parts of my life that had become an amusing anecdote I rolled out, or a haunting spectre I skimmed over. I had to find out where I had stored my deepest feelings and hold them up to the light to be looked at, not just by me, but by anyone who reads this book. To be frank, it was an idea that filled me with low-level dread.

My job was not just to recount my story in as interesting and as fun a way as possible, peppered with the usual failure, success and affirmations (of which there are several corkers), but to excavate and interrogate my responses, which I may have buried along the way. What did I really think about my childhood, my family, my career and myself? While I didn't see this book process as a form of therapy, what use would it be unless I was searingly honest?

I am a scientist. I deal in facts. And yet I can be just as enthralled by the mysteries of the universe as I am by the reality of the planet we live on, so here was my starting point, to straddle the two realms of information and emotion. I am also severely dyslexic. I read and I write in bite-sized chunks, so the chapters of this book contain short pieces and vignettes rather than long, rolling, dense text. It's how my brain works, just put down on the page.

A dear colleague of mine whom I used to work with referred to hindsight as his 'retrospectroscope', a made-up word which I fell in love with and has since become part of my lexicon. To me it conjures up the image of a powerful instrument that focuses its lens on past situations and looks at them retrospectively

THE RETROSPECTROSCOPE

with the light of today's knowledge. I have imagined packing one in my bag as I set out on this quest, time-travelling through the decades, beginning at a point just beyond my existence.

*

At the dawn of the Swinging Sixties, my father, Justus Adebayo Aderin, a tall, proud and driven man, left Nigeria and his teaching job in the hope of a better life for himself and his small family. Growing up in a colony country where the UK was considered the homeland, he thought England would be the best place to study medicine and qualify as a doctor. His sister had gone before him, and he'd gleaned sparse detail from her intermittent contact via airmail letters on gossamer-thin sheets of paper. She was the forerunner, and after much discussion with my mother it was seen as a good idea to transplant their little family and move them thousands of miles away from their relatives and community. While it was both a brave and naïve decision to take a leap into the unknown, Nigeria had just gained independence from British rule and there was escalating conflict. This probably spurred him on.

When Justus set sail, he left my mother, Caroline 'Bisi' Wey, and my eldest sister, Sue, a babe in arms, waving vigorously from the dock, with a plan to get settled before they joined him.

CHAPTER 1

On the long boat journey over, I am sure he hoped for a promised land which would receive him with open arms. Instead he arrived in the middle of a freezing winter without a coat (one had not been needed in Nigeria), to a country proudly displaying signs in guesthouse and pub windows saying 'No Blacks, No Dogs, No Irish'. The streets of London were not paved with potential, they were unwelcoming, hostile and icy cold. This was far from what he had hoped for, but with a dogged optimism which I think he passed on to me, he chose to make the most of the rather grim situation. So he invested in a warm coat, found jobs where he could and eventually ended up working at the Mount Pleasant sorting office. He also acquired a property on Pyrland Road in Islington.

I am not sure what my mum thought when she turned up several months later, but she didn't have time to dwell on any disappointment because she was soon pregnant with my second sister, Helen, who we call Hal. With a growing family and a belief in the UK education system, Mr and Mrs Aderin had to make the best of a bad situation.

One of the biggest disappointments of my father's life was that he never did get to study medicine. Instead, he had to find enough work to feed a family of four, which turned into five several years later. I was born at home on 9 March 1968 and was named Margaret Ebunoluwa Aderin, which was testimony to the dual existence I was about to embark on. Margaret was after Princess Margaret, who was cutting a dash through society at the time of my birth. My middle name is Yoruba (one of the main tribes in Nigeria and the one my people hail from) the word 'ebun' meaning gift and 'oluwa' meaning God. This was generous of my father because I was one of his other big disappointments when I turned out not to be a boy. Like many men of this time and before, I feel that he had a belief that to continue a legacy the proverbial baton had to be passed on via a male heir.

When I was very young I vowed that although I was a girl, I could be just as good as any boy; the realisation that I had let him down awoke a drive and competitive edge in me which has never left.

Then our family of five became six. One of my earliest memories is of my mum being pregnant. I was three and I can still recall the excitement and chatter around 'The Baby' coming. This new sibling seemed like the ideal playmate for me. So when I was told it had arrived, I was impatient to meet it.

I bounded up the staircase and into the living room of our home to see a small, wailing, scrunched up thing being washed in a large yellow baby bathtub which had handles at either end. I stopped abruptly. Mum was looking tired but was introducing me to 'Grace', which didn't make any sense to me. Where was The Baby? I didn't want Grace, I wanted The Baby, like I had been promised. I watched the water being carefully sloshed over this imposter and I started to cry.

CHAPTER 1

There had been complications with the birth and the next time I saw Grace was in hospital in an incubator. I stared at this tiny squirrely being, trapped in a glass tube, and felt the first stirrings of sibling protection and love.

Even though Gracie's birth is one of my earliest memories, while writing this memoir I have stumbled upon thoughts that are tenuous, fleeting – more feelings and emotions than concrete memories that have been long submerged in the deepest corners of my mind. One such memory was triggered by listening to a song.

I must have been two or three years old, as I am convinced that this happened before Gracie was born. I was lying in bed tossing and turning, unable to sleep. Sleeplessness has been a theme throughout my life, and I am sure that I will mention it later. On this occasion, I remember my mum was dressed up, possibly ready to go out for the evening with my father. My sisters were asleep but, as always, I was not. To soothe me into slumber, Mum would put a record on: *Scarlet Ribbons* sung by Harry Belafonte. In this memory, I knew she had played this to me before; it was somehow our song, and I loved the record and her warm, comforting presence. As it played, she hummed along, stroking my head getting me to relax. She was not in a hurry to leave or impatient with me, I had her full focus, and I was warm, content and happy. I was listening to the song, but I did not want to sleep because I knew she would go, and I loved having her company all to myself. But Harry's sweet tones lulled me into sleep, and there my memory ends. Today it is more a feeling that I am trying to make sense of, one that makes me sigh as I long for those simple times of contentment.

Years later I would get to meet Harry Belafonte as part of a project run by Yale University celebrating the 'out-of-the-box-thinking' of dyslexics. It turns out that Harry was not only a trailblazer in the music industry but also a fellow dyslexic thinker. I did not make the association at the time but just felt truly honoured to be sharing a stage with such a brilliant man. Later, I realised that his singing together with my mother's warm loving presence is my earliest and most formative memory.

*

My parents split up when I was four. Sue was eleven, Hal was nine and little Gracie was still counted in months, not years. I doubt I would have remembered much at this age if it hadn't been for what turned out to be the first of many custody battles that ultimately stretched into my teenage years. The court asked me who I wanted to live with, which is an impossible question to ask a small child who wants to be with both her parents and knows, even at that age, that whoever she chooses means she is rejecting the other. Children are far more aware of circumstances than they let on. Even as I was being reassured that whatever I said,

CHAPTER 1

I wouldn't offend anyone, I knew that not to be true. I sat in a chair, feet swinging, listening to the austere judge, and copied what my sisters or social worker had said, devolving all responsibility to avoid making up my own mind.

As I got older and this was repeated several times, I continued to feel as if I was stabbing someone in the back. It never got any easier. Sue and Hal were going through the same process as me, but they were older and, as such, deemed more culpable for their responses, and I can see that this must have made it even harder for them. It never got easier for any of us.

I met someone recently who told me how their husband had gone through a similar process, and it resulted in him being very conciliatory, not wanting to upset the apple cart and desperate to keep things as congenial as possible. I recognise this trait in myself, trying to make everyone happy, but I now know that this is not always possible.

During my parents' acrimonious divorce, I sought refuge in television programmes that transported me as far away as possible from my turbulent home life. *The Clangers*, an iconic stop-motion children's animation series, captured my imagination and heart, with pink mousey characters who lived on a moon-like planet, spoke in whistles, slurped green soup from the Soup Dragon and tucked into blue string pudding. If only I could escape these Earthly shackles and visit them in space! Space seemed to be a place that transcended terrestrial barriers; you don't see countries or borders from space, just our fabulous yet fragile planet. I imagined seeing our planet like that and what it would be like to go far beyond. This didn't seem impossible as everyone was still talking about Neil Armstrong landing on the moon, and I thought if he had gone out there, then why not me? This thought has never left me. *Why not me?*

When my parents broke up, I believe that my father moved to Avondale Crescent, in Brimsdown, a pocket of Enfield. Mum moved out of the house to be closer to her new beau, the vicar Barry Trill. My father, in what seems like a curious move considering their almighty fallout, had helped to set her up in a sweet shop with accommodation above.

I wanted to live with my father as I was often told by relatives and other folk that I was very much like him and this always made me very proud. When I was asked who I wanted to stay with, I just wanted to make a choice without hurting anyone, which was unfortunately impossible. So, although it mortifies me to this day, the idea of living above a sweet shop was a key deciding factor among many, in my agreement to live with Mum. I was incredibly close to my father, but I also loved sherbet lemons. What can I say? I was a very young child with a demanding sweet tooth. Also, although my opinion was sought, at four years old, I think that the court had its own ideas of where I should stay.

It made sense to all the grown-ups, except Father, that my sisters and I were split between him and Mum, so Sue and Hal stayed with him and Grace and I went to live with

CHAPTER 1

Mum. There were all sorts of accusations thrown around during this time. There was a rumour that Mum thought the older two had been influenced by Father and so it would make sense for them to stay with him. Father on the other hand said that he wouldn't stop fighting until he had all four of his daughters under his roof.

*

The first time I ran away from home, I was about five or six.

Gracie and I had settled in with Mum and most days I would pop downstairs to the shop to marvel at the large plastic bottles full of every kind of sweet imaginable. Mum was like the conductor of all things good. She arranged mounds of sweets in the shop window with bright colours and a defiantly hopeful order that was totally enticing. That order was a kind of expression of love. Years later, I read a book by one of my favourite authors, *Chocolat* by Joanne Harris. The main character was a brilliantly defiant woman who owned a chocolate shop, and she really reminded me of Mum. In our shop there was a long counter which ran around three sides of the space, and I would stand in the middle to stare at the shelves full of enticing treats and watch as Mum tipped them onto the weighing machine.

Images from this time are still quite hazy, like a fuzzy, jolty silent-film reel. There is one of me standing at the top of the long flight of stairs, waiting for Mum to come home, Gracie and me at home alone. Another is of my father, Sue and Hal coming to visit us at the sweet shop and finding Gracie and me there unaccompanied. I remember a snippet of conversation about possibly not living with Mum any more and going to live with Father and my sisters.

The snapshots I remember don't entirely marry with the emotions I experienced. I grasp hold of these nebulous feelings and the scattered thoughts, trying to work out what my reactions were at the time. The main emotions I remember are excitement while plotting something daring, the feeling of being treated like a grown-up and given responsibilities, which I knew were beyond someone my age. I think I may have felt guilty, too, knowing how my plans would affect my mother. I was helping to destabilise the conciliatory path I so wanted, but I am an uneasy narrator. The image coming through the retrospectroscope is dim and blurry and very emotionally charged, which makes it hard to analyse. One solution would be to turn to my family for their recollections, but if I do that then this will be their story and not mine.

I have always thought that the decision to run away was triggered by me, but this seems unlikely at such a young age. Maybe I was encouraged to believe it was my idea. I think at this stage of my life I was quite rambunctious and opinionated. I know my father wasn't happy that I may have been left home alone, in charge of an even younger sibling,

CHAPTER 1

although this wasn't an uncommon thing to do back then. If I had been like the majority of kids, I doubt I would have noticed that my mum went out at night, but here again the insomnia reared its head. Even when the family was all together, I would wander around the house while everyone else slept. I still find it hard to stop my thoughts and to settle down; I wonder if this played a factor in my becoming a stargazer? When you are up alone at night, the moon and stars can be a great comfort, and they have often helped me find peace when my mind is racing. Also, what better career than astronomy for someone who is up all night anyway?

A snippet of conversation from that time remains with me now, like smoke hanging in the air. 'It's not right. She's out partying with her new boyfriend and leaving the kids alone.' Is it a real memory and was it true?

On a number of visits to see Father and my older sisters, we talked about a plan that would bring Gracie and me back home where we belonged. Everything was building to what happened next.

I was given the responsibility of calling my father next time we were left alone. He would then come around with Sue and Hal for an impromptu visit, and finding us alone he would take us back to his house. My main memory is that I couldn't wait to use the phone. Something I was usually told not to play with was now something I could use; it felt so grown-up and sophisticated. The plan went ahead, with Sue and Hal coming in to get us, but Mum came home sooner than anticipated.

While they discussed why Sue and Hal were there, I slipped outside with Gracie. Which is how I came to be outside in the street at night hiding behind a bush dressed only in my favourite Womble-themed pyjamas and matching slippers. I knew it was wrong to be outside in my nightclothes and that I was likely to get in trouble for this, which was more worrying to me at the time than attempting to abscond. I was clutching Gracie's hand and begging her to keep as silent and still as a little mouse. She thought we were playing hide and seek and was more than willing to do as she was told. I had done everything that was expected of me and, as I was suspended midway between capture and escape, I could feel the excitement building at our daring adventure.

Once Sue and Hal left Mum they escorted us to Father's car which had been waiting nearby. We were bundled in and Father sped off. We had nothing with us. Packing a bag would have raised the alarm so we left everything behind. I think Father must have rung Mum when we got home to tell her we had decided to live with him and thus began another court battle. This time he was victorious and won full custody of the four of us, which was quite unusual for a man in the 1970s.

Through all of this I can't ever remember wanting my parents to get back together, but I did always want to make

CHAPTER 1

things better and for everybody to be happy and I think I felt guilty that this was often not the case.

*

My mum was a witch. Or at least that is what my father called her. While it was meant as an insult, he wasn't just using it as a pejorative female term but referring to a woman in Nigerian culture who is considered a malevolent, powerful, spiritual being. It was a word thick with fear and it was hard to process as it jarred with the memories I had of her singing me to sleep. But it did conjure up a mythical image of Mum that terrified me through my entire childhood.

I remember going around school thinking, *My mum's a witch!* and I even shouted it out one day when the teachers weren't around. There was a mixture of pride, horror and fascination in the belief that Mum had supernatural abilities, but my overriding memory was of being scared of her. It was a confusing time, to be living on the cusp of two very different worlds, in a clash of cultures, where at school I might be learning about the kings and queens of England and the Battle of Hastings, then at home I was being warned about the dangers of dark magic.

Father was keen to bring us up as pure and good, the inference, in his mind, being that he had to save us from what he feared were darker influences – fears rooted in the two worlds we lived in. We said our prayers religiously every night and, when I was older, we had an altar at home with a huge selection of religious paraphernalia. For a brief time, I was encouraged by my father to entertain the idea of being a nun, as he worried I was hypersensitive and would not cope in the cruel world.

My grandfather, whom I never met, was a minister in Nigeria, and many years later my uncle, who sadly passed away recently, was an archbishop there. Coupled with Mum starting a relationship with a vicar, we were surrounded by religion, and it shaped my childhood with prayers every night and the constant fear of doing something wrong. In fact, being scared felt like a normal emotion, not helped by my father's belief that there was evil everywhere. He constantly told us to be careful.

I remember Gracie and I attending Sunday school at a church. I was probably around seven and Gracie was three or four. The teacher wanted to teach all of us children the Lord's Prayer, but Gracie and I looked at each other in amazement and just started reciting it together. It seemed weird to be the only ones who knew it; didn't all kids say this with their parents before going to sleep at night? I am not sure what the teacher thought – hopefully that we were just very well brought up children.

As paranoid as Father seemed, he had experienced first-hand a dark undercurrent we had only glimpsed. He considered us to be strangers in a country we were regularly

CHAPTER 1

told to leave, and he felt the hostility keenly. I don't think he ever forgot the 'No Blacks, No Dogs, No Irish' signs he was faced with on his arrival in the UK, and this affected his ability to settle – which tainted us too.

Our home environment was a happy one but often an anxious one. In the safety of our own home we would dance around to the music of the time, listening to the radio or cassettes. It was Sue who brought Stevie Wonder into our lives and he was a firm favourite for all of us. We would also make up plays and generally lark around. But all of us were nervous of the outside world and the damage it could inflict. When the doorbell rang, for a time Sue would race upstairs to hide. Years later, as a teenager, I bought a second-hand Praktica single-lens reflex camera from a private seller. Having agreed a price on the phone when I was going to collect it, my father said, 'Be very careful when you knock on the door because they may not be expecting a Black person.'

Visits to Mum were court-mandated. She moved from the wonders of the sweet shop to a little house in Plaistow and then, when she married Barry the Vicar, to a big east London vicarage, where we would visit very occasionally.

Father told us to be careful of what we ate, in case it was laced with a spell that might influence us. Any gifts we were given were put into quarantine. Unsurprisingly, I was nervous about seeing Mum. I felt that I had to stay very close to God to be protected at all times. I think this was one of the reasons that I made a huge cross on one particular visit to the vicarage.

It was a sunny day, and I sat out on the concrete in the back garden, fashioning a cross from long pieces of wood with a hammer and nails. It was a good four metres high and towered over me. I managed to prop it up although I couldn't march around carrying it. That will protect me, I thought.

As for Mum having supernatural powers, I never did see much evidence of them when I was growing up. But the culture that my mother and father grew up in was very, very different from what I was experiencing in the UK. In West Africa, belief in dark magic is prevalent; what seems like paranoia in one environment is very much the norm in another.

*

I destroyed the kitchen twice at Avondale Crescent. The first time will sound odd, considering what I have just said about fear of the supernatural. I must have been around ten years old and we were watching the film *The Omen* as a family. There wasn't that much TV censorship at home when we were growing up, and I think that this film might have been seen as educational against the forces of darkness. Of course, I was way too young to be goggle-eyed at a horror film like

CHAPTER 1

this and hanging out with Damien, the demon child, but I thought I could handle it. In fact, I felt sorry for him. There is one scene where he discovers who he is and that he is evil through no fault of his own. This made me sad for him; it didn't seem fair that evil could be hereditary. Surely in a fair world evil must be a choice made by an individual, not just something that is thrust upon them? As the film goes on he starts to lean into the dark side and my empathy waned, but this isn't the point.

Whenever we watched films as a family, I loved taking the role of tea-maker, making strong, hot tea for everyone and for myself, which was super sweet. During *The Omen* I did the same but we didn't have an electric kettle, it was one that went on the gas hob, so I put it on and ran back to my seat so as not to miss too much of the film. I can't remember what the next scene was, but I was so freaked out by it that I was too scared to go back into the kitchen on my own. I was riveted to the spot. I knew if I asked someone else to go it would be admitting my immense fear, which would mean I would not be allowed to watch the end of the film, so I stayed quiet on the sofa while the kettle's plastic knob and handle slowly melted on the hob. By the time my blunder was discovered, the kitchen was filled with acrid smoke.

The second time I damaged the kitchen, it was down to pure forgetfulness. This is a long-standing theme in my life. I have only recently been diagnosed with ADHD but my mind has always been constantly abuzz with ideas, thoughts pinging about freely in my head. This can make me very distracted. I am doing one thing, then something else catches my eye and I am off doing that. Only the other day I picked some raspberries from the garden and put them in a pan to reduce them down to make a coulis. I went off to do something else, thinking about setting an alarm as a reminder, but this slipped my mind and I was sure I wouldn't forget. I came back to a burnt pot. In Avondale Crescent, it was more severe.

I loved to bathe; I still do. I was running the taps on full throttle at my father's house when I became distracted. The water overflowed and slopped onto the bathroom floor and still I was elsewhere, probably in my head contemplating the universe. The bathroom was above the kitchen, and the water worked its way through the floorboards and brought the ceiling below down. I don't remember my father being the slightest bit angry; things like that never seemed to annoy him. He accepted it as the accident it was.

My father's idea of punishment for Gracie and me was very different from the way he had been brought up in Nigeria, where he was likely to have got the stick. It was also not the way Sue and Hal were brought up and this caused some resentment. There were stark differences in discipline between the older and younger siblings in the family, but I didn't realise this until many years later. I have a memory of Father having two sticks: Mr White, which I think was a plastic curtain rod track, stiff with a bit of flex that made it a bit whip like; and Mr Brown, which was more solid

CHAPTER 1

and made of wood, rather like a broom handle. As far as my memory serves, I never interacted with the two rods of justice, but I know I feared them. I also know that Sue and Hal experienced the sharp sting of these two fellows for any major misdemeanours that they were caught doing. These two gentlemen seem very harsh, even brutal in the world we live in today, but in 1960s Nigeria I think that they were probably the norm. It is likely that both my mum and father had their own rods of justice while growing up. So it may be less surprising that my parents, who were still relatively young when they started a family, had transplanted these modes of discipline into our home.

By the time Gracie and I came along, my father had retired the rods and would instead use two fingers, smacking me on the palm of my hand, which barely hurt. But it was the humiliation that got me. After the smack I was made to stand in the corner. Often, I would lean my face against the wall, blubbing so hard that I left two wet patches behind. All that noise and fuss and I had barely been struck – this must have seemed to be quite an injustice to my sisters.

When we were older, during the holidays when we were all came home from our various schools, Sue and Hal would often be left alone and in charge of Gracie and me. Even then we were encouraged to study: the older ones revised for exams and Gracie and I were given sums and spellings.

Sue would teach me spellings, but she got frustrated with my lack of progress. I could never seem to get them right.

THE RETROSPECTROSCOPE

Now, I can see this is probably because of my dyslexia, but back then we had no idea about this. The implication was that I was stupid, lazy or not trying hard enough. On one occasion, when I got a spelling wrong, my sister Sue smacked me behind my knees with one of Father's slippers. I called them the pizza slippers because the pattern on the front of them reminded me of a margherita pizza with olives: red, yellow and black patches. This was nowhere near as bad as Misters White and Brown, but it did smart.

I was, once again, a snotty mess by the end of the lesson, crying at the shame and discomfort of it all. As I tried to walk away, I couldn't understand why my legs felt so strange. They had become swollen, so much so that they wouldn't bend. My sister was mortified and tried to bathe them with cold water in the hope the swelling would come down. Weeks later, my father found out about Sue's behaviour and was furious with her, which wouldn't have helped with her feelings of injustice. I know my sisters thought I was his favourite and told him so more than once.

Sue's hands could be strict, but the same hands could play music that was transformative. Over the school holidays when we were stuck at home and bored we would join her in her little box room where Sue would teach us songs that she had learned, singing softly while strumming. The songs she taught us are still with me today, I can hear them ringing out from somewhere deep in my memories. Some were upbeat and religious in character, like:

CHAPTER 1

Where, tell me, where can I find you
Can I find you in a fast-flowing stream,
Can I find you in my wife,
You know I need you in my life,
Oh, please won't you tell me where you are.
I'm gonna look for you when I don't really want to
I'm gonna look for you when I'm tired and I'm worn
I'm gonna look along the way it's a lost and lonely day
Like a morning waiting for a dawn

Some were darker, like 'I Wish We'd All Been Ready' by Larry Norman.

I did not realise this at the time, but I recently looked up this last song and found out it comes from *A Thief in the Night* film series, which depicts a Christian event called The Rapture, where the god-fearing are taken to heaven and the rest are left behind. I guess the fear was with us always.

After we had left Mum's, I remember my father was working all the time to support us, often with two or three jobs on the go.

I also felt that my sisters wanted to toughen me up. After my departure from my mother's care I think I lost some of my spark. Before this I had been an overconfident child who would take on most things, but this shifted and I became a bit of a wimp. I would cry often. They would push me into a patch of grass in the garden that was alive with crickets. I was so scared of them as they jumped up around my legs.

I was also the sickly one of the family, the only one who had asthma and terrible eczema; I couldn't stop scratching. As a result my whole family called me 'Margaret Scratcher' (after our PM at the time – I was initially proud to share a name with her, but the shine soon wore off). Even my father joined in. I think they thought I wasn't trying hard enough not to scratch, and with better willpower I would be able to stop. Both of my conditions are exacerbated by stress, so the name-calling really did not help.

Today, I have an incredible radar for bullying; I find it really hard to watch programmes where someone is getting picked on or bullied. It is triggering and makes my hackles rise, and I have a physical response to them. If I see controlling behaviour, I get very upset and I want to defend the underdog. I feel a sense of disloyalty including these stories, but I want to because these incidents have made me who I am. I also know I did my fair share of goading and was not always the innocent party. I have very embarrassing memories of me, not long after being pot-trained, standing at the bottom of the stairs and needing the loo so badly I could not walk, and my parents making Sue or Hal carry me up the stairs. We can all rewrite history, particularly where our siblings are concerned. Long-running feuds, hierarchical control issues and sheer bloody-mindedness can echo around families and gradually calcify as we age. My

CHAPTER 1

family has suffered from that. Although we have been very close at times, many unresolved issues remain.

As an adult, I can see how hard it must have been for teenage girls to have to look after their younger siblings. They didn't have a choice. I didn't realise what they were going through because I was too young to make sense of it or understand that this wasn't how other families operated.

I pined for my father when he was at work; I knew things would be more settled when he got home. As an insomniac, I would sometimes sneak downstairs and find my father exhausted, sleeping on the sofa. I would often lie down next to him and rest my head on his chest so I could hear his heartbeat. It made me feel so content. Even today, thinking back on the memory, I consider it one of the happiest times of my life. Lying warm and safe in his arms. The resonance of his voice through his chest when he spoke, and his soft heartbeat. But it couldn't last. My sisters pointed out to me that it was selfish to be disturbing Father after all the hours he put in, and I knew that they were right; I stayed in my bed after that.

*

Father and I were very similar, and I loved his company. One day, I asked him if he had any photographs of himself as a child.

THE RETROSPECTROSCOPE

'Oh yes,' he said, 'come upstairs and I'll show you.' I bounded after him, excited to see pictures I had never seen before. We went into the bedroom, and he stood me in front of the mirror. 'That's what I looked like,' he was smiling at my reflection. 'Just like you.' I was so proud, I could have burst right there and then. There was nobody I would rather have looked like. At one point I tried to shave my hair to match my father's slightly receding hair line – that's how much I wanted to be like him. We all knew that he was working all hours to keep us going and we loved him deeply for the sacrifices he was making.

Through all of this I thought that there was one thing that could make Father proud. I was determined to be a worthy boy substitute, but I wasn't pretending to share Father's interests to garner favour. While I had inherited the passion for science that ran through his veins, it was my own fascination with space that fuelled my curiosity.

I was introduced to the moon by my father at a very early age. He told me of his love of its beauty and how it had been his friend as he cycled home from school on his Raleigh bicycle. It was a 12-mile journey for him to get to school and at night the moon was his friend as it would light his way home. He described himself as a 'self-certified lunatic' and would emphasise the importance of the self-certification. I later adopted the term myself, figuring if the moon was my father's friend then it was mine too. It turns out that lunacy is hereditary and I have passed on this love and fascination

CHAPTER 1

of the moon to my daughter Lori. On clear moonlit nights we sometimes step outside and howl at the moon together. It is cathartic, primal and a really good laugh. I am not sure what our neighbours think about it, though.

My heroes were very logical characters – Spock from *Star Trek* and Arthur Conan Doyle's Sherlock Holmes – both of whom could successfully control their emotions, remaining clinical, scientific and analytical no matter what the circumstance. This really appealed to me, as I often felt overwhelmed by my emotions. I felt *too much* and I would lie awake at night, imagining what it would be like if I could just turn the feelings off. My experience of grown-ups was fraught with chaos, drama and conflict, so I sought out role models who transcended the mess of emotions to reside on a higher plane. It was an antidote to my overwhelming sensitivity to the world around me, something my mum first picked up on when I was just a baby. She noticed I was a considerate breastfeeder. She said I was the only one of her children who could tell if she was in pain or discomfort, if she winced I would pull away and wait for her to get comfortable before latching on again. When she told me of this memory recently it made us both tearful.

As I grew older, this became more evident – not just where my family was concerned, but also on random issues and inanimate objects. Like the time I sat at the breakfast table in silent tears.

'What on earth are you crying about now?' one of my older sisters asked, exasperated by my unexpected emotional outbursts.

'I had toast,' I pointed at the crumbs on my plate, 'and I feel bad. Maybe the bread wanted me. I made a choice between them and now I am worried that the bread might be upset.' A psychotherapist might connect this reaction to the trauma of having to choose between my parents. I am not saying this is wrong, but I think my behaviour also comes from an inherently empathetic nature and from not knowing how to stop or moderate it. Back then I can see it probably seemed sort of deranged. Yet even now, I have a sensitivity towards inanimate objects; if my daughter abandons a soft toy in the house, I will always put it in a more comfortable position. They just look happier that way.

It makes me wonder whether empathy is learned or inherent? This is one of the subjects we talk about in my work as an ambassador for the charity Made By Dyslexia. Research has shown that dyslexics have a stronger emotional intelligence then non-neurodiverse people and a highly developed understanding of people and situations, which can make them more empathetic and focused on justice – a powerful tool to have in today's world. This has often been my experience as a dyslexic, but can empathy be learned, grown or enhanced? Is there such a thing as too much empathy? Answers on a postcard, please!

*

CHAPTER 1

I went to thirteen schools in twelve years. Perhaps that is some sort of record. My educational history is bundled up in one big jumble of uniforms, playgrounds and people. Once I was back under my father's roof, I was sent to Brimsdown Primary School in north London with my older sisters. This was the third primary school I had attended. We would have to cross a bridge over a train track and, if a train was passing underneath, I would freak out. I was absolutely terrified, convinced the vibrations would cause the bridge to collapse. Sometimes my sisters would think it was funny to run away and leave me up there when the train was coming.

I was in the lower school; my sisters were in the upper school and our playgrounds were separated by a wire fence. Even though we did not always get on, I had terrible separation anxiety and was desperate not to be parted from them. I would hang on to the fence for dear life, wrapping my tiny fingers around the wire, trying to stay with them for as long as possible. Many years later I saw one of the X-Men films, in which a young Magneto could control metal with his mind, but only when under duress. Although the circumstances were very different, when he was separated from his Mum, he warped the fence. It looked so familiar that I could imagine myself doing the same and stepping through to my sisters. At the time it took two teachers to prise me off fence and drag me to my classroom.

It may sound like I hated school, but I didn't. Or at least, I was happy at break time when I could join a game of tag and run the length of the playground. I would take off, legs effortlessly stretching out, every stride covering more concrete until I would look back and think, *Hey, look how far I have come!* I wasn't even out of breath. It felt like I was flying. (I always loved running and had to stop recently because of an injury, but my knees are getting better, so I am really hoping I will be allowed to run again soon; I've really missed it. Although I don't fly like I did when I was young.)

I was not as fond of being in the classroom. Even at the age of six, I could sense a level of judgement around my capabilities so I would often pretend to fall asleep at my desk. Sometimes, this turned into an actual snooze. I found reading hard and didn't graduate from the easy red books to the blue books, the colours singling out my hopelessness as other pupils steadily progressed through the levels. To coin a phrase from Harry Enfield, I was the 'nice, but dim' child who spent a lot of time at the back of the class with a pair of safety scissors and the glue. Creating things kept me occupied and happy.

For one art project, we had to make a dragon. Oh, how I loved dragons! This assignment was made for me. We were all given a section to paint and mine was a tiny piece which I had to fill with yellow. That was it. No scales on the body, ferocious fire breathing or whip-smart tail, just a single block of colour. I found this disappointing because I wanted to use my artistic prowess. I also wanted the experience to last for as long as possible so, once I had painted

CHAPTER 1

the paper yellow, I painted it again and then again. If I was going to be given a small task, I was going to make the most of it and I planned to use as much paint as if I had painted the whole dragon.

At Christmas there was a Nativity play. We were all on stage to sing a song about Christmas presents and the cool, clever kids got to be at the front. Each of them had a letter pinned to their back and, as they turned around to face the back of the stage, it spelled out CHRISTMAS. I was stuck at the back and not singled out to be a letter bearer, much to my chagrin. The thespian in me was desperate to shine. I sang the song with gusto and can still remember it now, which is a bit unsettling considering the number of things I have forgotten about my past. Or what I am supposed to be doing tomorrow. Yet, this tune lives rent-free in my head.

My teacher told my father that they were going to keep me back a year because I wasn't 'up to snuff'. She must have thought I had a lot going against me: a Black kid from a broken home being cared for by a hard-working but regularly absent father. I had a number of labels attached to me, the biggest – and most annoying – of which seemed to be that I was stupid. At the time, barely anybody had heard of dyslexia. Certainly, not me. Frustratingly, I was placed in remedial classes and it felt like the system had already given up on me – but one person didn't: my father repeatedly told me how clever I was.

THE RETROSPECTROSCOPE

From the age of four, Father would ask me, 'So which Oxbridge college are you going to, Maggie?' In his mind, and typical of many Nigerians I have met, he valued education above most other things and he had high expectations for each of his daughters. In this he was progressive, knowing it was our ticket out of a toxic environment which looked down on women and Black people. We were the double whammy. He wanted one of us to be a lawyer, one to be an accountant and he was adamant that I would go into medicine. Father's mantra was, if we worked hard enough, we would be amazed at what we could achieve. He didn't just want us to survive, he wanted us to thrive, so he pushed us determinedly through our educational journeys, and I am so glad he did.

*

Identity is a fascinating thing to me. For as long as I can remember, mine has been in question, either by others or, most importantly, by me.

As a child, I defined myself very much as a Nigerian, even though I had never been there. If I said I was British I was worried someone might question this. I had that experience in the playground at school; kids pointed at me and said, 'You're not from here, you're the wrong colour, why don't you go home?' My initial thought was that I only

CHAPTER 1

lived around the corner, and they must be confused. When I realised they meant leaving the UK, where I was born and had never left, I thought, *Back home to where?* To Nigeria? How could a place I had never been to be home?

I now realise that this was not the kids talking but their parents, but it still left me feeling like I didn't belong, so I made an active decision not to be British but to be Nigerian, just as my proud father and mother were, and this seemed a good solution.

When some of my classmates treated me as if I didn't belong, I didn't persuade them otherwise because I had my pride; why try to be something that others didn't want me to be? That seemed like a recipe for trouble, particularly where the boys were concerned, because they teased me a lot, which made me incredibly wary of them. Throughout my early years I was either the only Black kid in the class or one of very few – at every school I went to.

I embraced my heritage; it was important to me that I was Nigerian. In the middle of the heatwave in 1976, I strutted around in a thick jumper, declaring that I could handle hot weather because I was Nigerian, until I reached sweltering point and had to admit defeat. It appeared I had to prove who I was, not only to others but to myself as well.

I hadn't been to Nigeria, and I had no idea what it was like, but when Gracie was born, she was wrapped in a piece of Nigerian material, and this represented the entire country to me. The fabric was dark and shot through with interesting oranges, browns and black, and it was the only tangible thing I had to connect me to the place.

The problem was that while I didn't feel comfortable calling the UK my home, I didn't fit into the Nigerian culture either. We would meet up with relatives who had come over from Nigeria to visit, but I didn't speak Yoruba. They would shake their heads sadly and say, 'You're a lost Nigerian, you don't speak the language, you've never been there.'

Although both my parents were born in Nigeria, they spent the largest part of their lives in the UK and Mum, the granddaughter of a well-to-do admiral, is incredibly British in her behaviour, with a clipped accent to match. For many years it was confusing to me: where I was from, who it made me, what I was allowed to be and why I was waiting for permission to be one thing or the other. I was searching for an identity that seemed impossible to pin down. I realised, rather late on in my life, why not be proudly both?

And knowing that was a relief, because I could embrace all the wonderful parts of my mixed heritage.

*

Gracie and I settled into our new life with our father and sisters in the house on Avondale Crescent. It must have been tough for him with two teenagers, one at primary school (me) and a toddler. That's a big spread of ages and emotional

CHAPTER 1

needs to juggle. Father worked full time and for a while we muddled along with various after-school arrangements, with my older sisters becoming unofficial carers for their younger siblings. This can't have been easy for them.

Some of the time I was a latchkey kid. If there was nobody in when I got home, I knew I could pop over the road to the Nidos, an Asian family who lived on the corner and were happy for me to hang out until one of my older sisters came home. Mr Nido had had a stroke and I was a little scared of him, but they were all very kind to me.

If I was feeling brave enough, I would let myself into our house. I can visualise turning the key in the lock and coming through the front door into an empty, silent home. There was a narrow hallway with stairs ahead of me leading up to three bedrooms and immediately to the left was the living room with a sofa and an enormous radiogram: a radio and record player combined. I never touched it, but Father would play music in the evening. Our TV must have been in there too, black and white back then of course.

There was a room divider which separated the living area from the dining table at the other end. We weren't a sitting-down-at-the-table sort of family, so I am not sure why the room was divided in that way. The kitchen was off to the side and led back through to the hallway.

Upstairs, Sue had the box room. I shared the double bed in the back bedroom with Gracie and when Hal was with us, in a single bed. Sometimes, when we had a live-in housekeeper, Father would also sleep in there. And Simon, of course. Simon was our cat who I adored even though I was terribly allergic, not helped by the fact that I encouraged him to sleep at the end of my bed.

My auntie was quite prominent in our early lives, taking a female carer role when my father was working. I dreaded going to stay with her for three main reasons. Firstly, we were not allowed a bath. We had to get a bucket of water and a cup and then stand in the bath, dousing ourselves, which was a nod to Nigerian culture and also saved money. Secondly, she brushed our hair and by God did it hurt. I go to an African hairdresser now and can withstand all the tugging and pulling because of my auntie's brutal approach to my resilient hair. She once broke the teeth from the comb as she dragged it through my unruly tresses and tutted at me not to complain as silent tears ran down my face. It didn't take long for me to learn how to do my own plaits! Thirdly, she fed us a Nigerian breakfast called Ogi, traditionally made from maize which has been soaked and ground before adding hot water. Auntie used cornflour instead, which made a gloopy slop and was almost impossible to stomach, but she made us eat it. It was like instant custard without the colour, flavour or texture. Those visits were always associated with discomfort.

As penny-pinching as my auntie could be, my father was not like that. We used to go shopping at the frozen-food supermarket, Bejam (before it became Iceland) and he would tell me to choose what I wanted. One of my earliest

CHAPTER 1

memories is opening a chest freezer we had and balancing the door on my head so I could reach in for the ice pops below, a firm favourite of mine. The food we bought was put in a very large paper sack and, when we got home, I would empty it and cut out three holes for my arms and face and wander around pretending to be a robot.

Some of my cousins lived in Nigeria because my auntie had all her children there before she came to the UK, and they would come over to stay. I spent a lot of time hanging out with my cousin Remi; one day, I was told that he had been deported. This was a completely alien idea to me. It seemed odd to me that a relative could be thrown out of the country I called my home and where his family was based. I wondered whether my footing here wasn't quite as secure as I thought it was and it unsettled me. It didn't take much to make me feel anxious.

As well as Auntie, we had a succession of women housekeepers who would look after us. Occasionally one of them would move in for a while, like Mrs McCluskey, who turned up with her son. She was Scottish and her ways were a little unfamiliar to us. She once asked me to wash the salad before supper and, copying how we cleaned our vegetables in hot water, I caused the leaves to wilt beyond salvation. Her cooking coupled with our school lunches made me realise how Nigerian we were in terms of what we ate.

School was a stressful place for me, but one of the biggest challenges was lunchtime. In the queue in the school canteen, I would clutch my tray and hold it out to the dinner lady who slopped something unrecognisable onto my plate. The first time I had cabbage, I thought, *Good God, what is this?!* It looked like they had just popped out to the garden to pick some unidentifiable leaves. I was used to vegetables being incorporated in a dish, rather than overboiled, soggy and leaking water onto my plate.

At home, one of our staple dishes was the popular West African dish jollof rice, which is basmati rice cooked with tomatoes, onions, peppers and spices and served with chicken, lamb or beef. I still cook it now, along with another family favourite of finely ground rice which is mixed with hot water to make a dough. I put a generous dollop on to the plate and make a shallow dip in the middle to pour in a tomato stew with meaty bits and vegetables, served with boiled okra. Like many cultures, Nigerians often eat with their hands, so we break off chunks of the rice dough and dunk it into the delicious stew.

My sweet tooth was encouraged by my beloved childhood pudding of garri, a flour made from pounded cassava, which was soaked in water and then sprinkled with sugar. Once I was old enough to be trusted, I was allowed to add my own sugar to my bowl, but this was a rookie move on behalf of my father because I just ramped up the sweetness. I was also addicted to Sugar Puffs for breakfast. The Honey Monster was always welcome at our table. It's a wonder I have any teeth left.

CHAPTER 1

For a while, we ate a lot of pizza. Father had been supplying dough and other ingredients to all the Pizza Express restaurants in London. He did this from their industrial kitchen on Arlington Road in Camden Town. Sometimes when he had to work late to fulfil an order, he took us back there after school. I loved this place with its robotic metal machines. Upstairs was a bit of a dumping ground, and there was a very old-fashioned till that Gracie and I loved messing around with, pressing buttons and watching numbers pop up but in old money. Father would be downstairs making the pizza dough in what looked like a giant cauldron. He poured the flour and yeast into the mixer, set it going and we watched, transfixed by the dough-mixing legs that would jump up and down, kneading the ingredients. Once the dough was made, it was cut, rolled into balls and stored in the freezer ready to be sent out to various restaurants across London. Every so often, he would pause and make us a pizza and we would beg for extra pepperoni.

At least in one home-cooked meal we were aligned with most other children across the UK: beans on toast. They divided our family though, because they would be my choice but Gracie preferred spaghetti hoops. I hated them and she detested beans, so we would provoke each other over our chosen teatime meal. This was a rare disagreement because in most other things we were the best of friends.

*

THE RETROSPECTROSCOPE

I was so very close to Gracie. She was my first audience and my safe harbour; in looking after her I learned the shape of my own caring.

I first learned how to plait hair so I could do hers regularly, thereby avoiding our auntie's hairdressing. To begin with I was told off because it was thought that I would create knots in Gracie's hair, so I would do it at night when there was no interference. Having done a good job I was allowed to regularly plait her hair. I guess I was six or seven when I started doing this. I became more of a mini-mum to her than an older sister, and she responded positively to that. It felt good to look after someone younger than me, and a relief after being dictated to by my older sisters. I sensed Gracie's vulnerability and wanted to protect her.

Gracie was a distraction for me throughout the chaos of our upbringing and gave me something to focus on besides myself, which helped a lot. I knew if she was OK then I would be all right too, which set the tone for our future relationship; when Father died many years later, my main priority was Gracie. It may sound caring and generous, but it wasn't really selfless because worrying about her gave me a break from my own grief. 'I'm OK,' I would say. 'It's the baby of the family I am worried about.'

I have always been a natural storyteller and Gracie was my first willing listener. I was forever thinking of tales to entertain her and creating imaginary worlds for us to play in. Often, she was Princess Grace, and I was the wicked villain

CHAPTER 1

rather unimaginatively called 'Wicked'. In this role I had a pencil-thin moustache of course, a cape and a dastardly plan. Gracie loved our pretend adventures and being the princess, and I loved being everybody else: a multitude of interesting characters, some good, some evil. That was my job.

We couldn't go anywhere without me conjuring up a story around it, whether it was about the Shell figure in the logo when we pulled into the petrol station or the traffic around us. For some reason we called big vehicles 'bolly boots' and there were good bolly boots like buses because they carried people around and bad bolly boots like trucks because they seemed scary to us and could crush us under their monster tyres.

Creating imaginary scenarios was my happy place. I really enjoyed finding ways for Gracie and me to escape the turbulence of our existence. We had our own little world we could slip off to, where we were safe and in control, which meant whatever was happening out there in our real lives didn't affect us so much.

There was soon another school move, this time to Chase Side Primary. Or it might have been Raynham Primary – with so many schools, it is hard to get the order right. But, at eight, I was sent to boarding school. And Gracie? Well, she was sent to Nigeria. It felt as if we had been physically wrenched apart.

Chapter 2
Boarding School Blues

I was interviewed by the *Observer* recently. To illustrate the article, they wanted a childhood photograph of me that we would then recreate in the present day. For some interviewees, it was an old picture of them playing with dolls or a train set, but I had very few toys and even fewer photographs to prove it. I gave them my passport photo.

I must have been about seven and I had been taken to a photographer's shop in Camden Town with Gracie. We were getting snapped for our Nigerian passports. At this point, I think the plan was to send us both to Nigeria because Father was working full time and struggling to look after us all.

We were wearing matching dresses, in different colours, which we had pointed out to Father in Woolworths. They each had a collar with flowers on; mine was green and Gracie's was red or orange. We had to sit up straight – it was quite a formal affair in the studio. The photographer fussed around us and sang a made-up ditty with the words,

CHAPTER 2

'How can I make you look natural?' It became a mantra in our household, which is why I can still remember it. Sitting there in our official passport poses, the answer was that it was not going to be easy to make us look natural – and the singing was certainly not helping.

In the end, I didn't have to use my passport because I was sent to boarding school instead. My clever sister, Hal, had set the trend when she got a scholarship, meaning reduced fees, to a school called Christ's Hospital. She was the sibling I idolised the most and wanted to be like in every way. I used to try and write like Hal and copied her French style with numbers, which meant crossing my sevens. This backfired because, as a dyslexic, I would easily get my numerals muddled – for example, how 4 and 7 were formed via similar strokes of the pen.

Hal was – and still is – wise, sophisticated and confident, with a sharp wit, which I often cut myself on. I, on the other hand, saw myself as dumb and a little timid. My pre-school confidence had waned. It wasn't just leaving Mum and the sweet shop that had changed me; I think those initial years at school taught me I was not as bright as everyone else. Today, I love the fact that I am dyslexic and truly see the amazing traits it gives me, especially in my work as a science communicator. But when you first get to school it is all about reading and writing, and these were the things that I found particularly hard. I was transformed from a confident toddler who was thought to be quite bright, to an introverted student who was placed at the back of the class and kept occupied with safety scissors and glue.

Hal's school looked more like a prison than a place of education. It was an imposing building with large wrought-iron gates and two solemn women in old-fashioned garb standing atop the gate posts. As we dropped her off it looked as if she would be doing hard time, smashing rocks or walking around with a ball and chain.

During Hal's time the students were split into houses and the younger years, of which Hal was one, had to sit at the far end of the long refectory tables. There was a hierarchy for serving food which started with the oldest in the school. By the time the dishes reached her there were just scrapings left. Hal lost a considerable amount of weight during those first months, so Father sent her back at the beginning of each new term with a tuck box of food which he had carefully and lovingly packed to see her through.

I don't think Hal had a good time at boarding school, or at least not initially. She was used to being a big fish in a small pond at home, with two younger adoring sisters, and now she was at the bottom of the pecking order. This is often the way when pupils start secondary school.

Sue was still at home in the early days of Hal being absent, but Father felt it made sense for her to go away to school too. There was no scholarship this time around so Sue went to Fernhill Manor School in Hampshire. Soon after, with two of us left at home, my father asked me if

CHAPTER 2

I wanted to go to boarding school and I jumped at the chance; it seemed like the ultimate adventure. I pushed all the negatives of imposing buildings, tricky student dynamics and a lack of food to the back of my mind and decided it looked like a really exciting thing to do. I guess I wanted to follow in the steps of my older sisters, too. I wondered what treats would be packed in my tuck box.

Except I wasn't nearly as clever as Hal, who loved reading, so there was no opportunity for a scholarship. Instead, Father selected several possible schools and took me for a series of first interviews with a range of headteachers. Father and I would sit in a smart study, making small talk while sipping tea from fine bone china cups. I loved tea, or at least I did the way I made it, which was very strong and sweet. At one school I accepted the offer of a cuppa and then proceeded to heap five sugars into my cup, stirring it vigorously and then diligently licking the spoon. (I didn't want to waste any of that precious syrup.) The adults looked on aghast and I noticed a frown on my father's face. When we got home my father didn't tell me off for my bad manners, but suggested next time it might be best if I didn't put the spoon in my mouth.

Amazingly, I got an offer from a rather smart prep school, Ladymede in Buckinghamshire, but I was quite young so my father and I had another discussion about whether boarding was really right for me, especially given how sensitive a creature I was. My sister had started boarding at secondary school level, whereas I was only eight. On my insistence we accepted the place. Many years later, my father talked about regretting this decision, not just for me but for all my sisters, when he discovered there was a very good girls' school down the road from where he worked in Camden. We could have attended as day pupils and slept in our own beds at night. I know he hated us being away, but he was working every hour and all but one of us chose our fate.

As I was packing to go to boarding school, four-year-old Gracie flew to Nigeria to stay with our father's relatives. I hoped that she wouldn't miss home too badly, but she hated being there. Not only was the environment a complete contrast to the UK, but for a child it was markedly different because, in Nigeria at that time, it was acceptable to physically punish children. She must have been introduced to the equivalent of Misters White and Brown. The cultural change must have been overwhelming, and she was miserable.

I was also bereft without Gracie. We were so close, and I had taken my role of surrogate mum very seriously. Her absence was compounded by the loss I had already experienced through my parents' break-up, followed by my older sisters going away to school. I think this spurred me on to embrace the idea of leaving too, yet on the day of my departure, I hated saying goodbye to my father and knew I would miss him terribly.

He drove me to school and, as we left Central London, I kept my eye on what was then called the Post Office Tower

CHAPTER 2

for as long as possible, until it was out of sight. It was renamed the BT Tower in the 1980s. This became a familiar marker in my future journeys, as did another similar tower on the M40, which signalled school was ahead. We didn't take this route very often because, for a couple of years, I was a 'lifer', only coming back for the Christmas and summer holidays. As an adult, I was invited to a shindig in the rotating restaurant at the top of the BT Tower and it was quite a moment for me to stand in such an emotive building, a place that represented home when I spotted it on the skyline.

*

Ladymede Preparatory School in Aylesbury was quite posh, with a smattering of girls from eminent families. I joined around the same time as Lord Mountbatten's granddaughter India Hicks, and I remember the tragedy of his death when the IRA bombed his boat in Ireland in 1979. It happened during our summer break and, when we all returned, we were told to be very sympathetic. I wasn't close to India, particularly not after she had 'stolen' my best friend, Louise Osmond, but we didn't need to be reminded to be kind because that was already the culture of the school and we all felt really bad for her.

Life at Ladymede was a world away from what I was used to. I was the only Black girl there and I had no important family history or social standing, yet I didn't feel like an outsider. The school made it clear to me they had a zero-tolerance attitude to racism, and I was to report this behaviour immediately. There was one silly comment from another girl, and she was spoken to, but that was the sum total of it.

It's funny because in my eyes, I just saw a uniform group of people, but everyone else saw that uniformed group of people *and me*, the Black kid. My difference didn't stand out to me, and I didn't consider myself separate. I think I am more aware of this now than I was in the past, maybe because of the industry I am in. I notice when I am the only woman in the room, or the only Black person. Sometimes, if I am driving around an unfamiliar town, off to speak at a literary festival maybe, I am suddenly aware that I have not seen another Black person. I can sit on stage for an event and look out at a sea of white faces.

This is an observation I make now which I never did when I was growing up. I don't see these situations as a problem. To my mind, there are a multiplicity of things I may have in common with an audience – fellow lunatics, a deep and abiding love of the stars and countless other potential overlaps – so the colour of my skin seems irrelevant compared to all the things we may share. But in some circumstances it does add a certain amount of pressure, as it can feel like I am an ambassador for my race, especially when I am abroad.

I was not the only student at Ladymede with a culturally diverse heritage. Several girls had come over from different

CHAPTER 2

countries to attend the school and some had parents who lived overseas. Deborah Carr was a marvellous student, universally admired because of her sporting prowess and natural beauty. She came from an island with an active volcano that erupted once while we were at school. She was remarkably calm about the whole thing and luckily her family were unscathed, but this gives an idea of the global reach of some of the girls. Having lived such a relatively sheltered life, I enjoyed meeting and sharing thoughts with such a wide range of people. Sara Jane Arida, for example, came from Lebanon and her father was important in politics there, but it was a country I had never heard of before I got to Ladymede.

I settled quickly, helped by the fact that it was a small school, and my peers were very welcoming. They would occasionally take me in hand to help me fit in, probably more than I realised at the time, but it was never done with any agenda or malice. Like the time I said 'crap', which was a word we always used at home, but when I said it at school for the first time, some of the girls looked at me in horror.

'Margaret!' they said in unison, 'you really mustn't say that! It's not a polite word.' It was news to me. At home we didn't swear, but 'crap' was common parlance. I made sure I refrained from using it under their tutelage.

I also pronounced 'ask' as 'aks', putting the 'k' before the 's' which is a very Nigerian way of saying it. The dyslexia confused things further, so even when it was pointed out to me I couldn't see the difference.

'It's ask, as in A S K,' one of my friends would say.

'Yes, that's what I said, A S K, aks,' I replied. It took a long time for me to first hear the difference and then correct myself, and I would often slip back into it.

The way I speak today is a direct product of some of the schools I went to, but also my big sister, Hal. If I was trying to emulate anyone it would have been her and her boarding school timbre, because I idolised her. I was not aware of my voice changing, or putting on an accent, but it did and now it would be almost impossible to detect that I am the daughter of Nigerian parents. The boarding school experience carved me into a girl with middle-class vowels and mannerisms – although I still say 'aks' and generally care not a jot. Recently I was doing a podcast for the BBC World Service and the producer, Florian, noticed that it still creeps in. English was not his first language, so he was quite sympathetic about it.

I didn't choose to sound different from where I started and I was not aware of any stigma attached to it, but I may have subconsciously changed my accent to help myself acclimatise. I went to so many different schools that I felt I could fit in anywhere. Rather than floundering, I chose to power through and adapt to each new experience. I shifted shape and blended in without realising this was what I was doing, finding a comfortable spot to rest before I moved on again.

At Ladymede, it was more than just my speech that was transforming: my friends were teaching me things they

CHAPTER 2

had grown up doing, like playing tennis. I had never been on a tennis court before or held a racquet in my hand, so a group of pals took me for a knockabout. After the game, we were all chatting, leaning on the net, and the wire holding it up snapped. We were convinced we would get into terrible trouble and scarpered. I don't recall if we were ever caught for the misdemeanour, but I do remember the feeling of mortification that came with it. That's difficult to shake.

Sports day was an important date in the calendar. We would be dressed in our big, baggy, pale blue PE pants, self-consciously pulled up over our normal pants. Imagine running around and jumping in a pair of unwieldy brushed-cotton bloomers! When all the parents arrived, we stood in a long line and had to perform a strange sort of cheerleader's welcome, by falling gracefully to the ground one by one. It was a synchronised cascade of girls, as we each tipped forward, raising one leg behind us as we fell, arms outstretched to break our descent. Then we all sat up on our haunches and the competition could begin.

I became a demon croquet player because there was a pitch on the front lawn of the school, and I took part in a tournament, teamed with Deborah. I think that many people felt sorry for her being lumbered with me, a complete novice, but I held my own and learned the game fast. I believe that we got to the final and may have even won. Memories are hazy on this.

BOARDING SCHOOL BLUES

The hula hoop craze swept around the school. We had competitions and I wasn't bad, but I wasn't a patch on Deborah, who could keep the hoop moving for a couple of hours. My thing was maypole dancing, I was a dab hand at it. I loved weaving intricate patterns with the ribbon on the pole and then unravelling them, but it was the dancing I adored the most. At one of my schools, we learned Scottish dancing, and this was by far my favourite, with the coordination, teamwork and beautifully choreographed end result. My daughter, Lori, and I were watching *Bridgerton* recently and during one of the spectacular balls, there was a formal dance which reminded me of these dances we did at school. Lori thought it was hilarious and it furthered her belief that I was brought up in Dickensian times.

Growing up I never seemed to have the appropriate equipment for sports or the right tools for any job, but I loved to improvise. From using a knife as a screwdriver to change a fuse or using a full-to-the-brim bottle as a hammer, the correct tool was often just not available – sometimes because we couldn't afford it, sometimes because it had been misplaced in the constant house moving. My husband, Martin, used to find this very strange as his father was a car mechanic, so a wide variety of the right tools were always available to him. When he saw me doing my knife screwdriver trick for the first time, I think that he thought I was mad.

An example of this in action was the time one of my boarding school friends gave me a geode from Bahrain,

CHAPTER 2

where she lived during the holidays. She got it for me because I was fascinated by rocks and crystals. I wanted to open it, but I didn't have anything to help me, so I tried using a fork, then the bottle hammer with a knife. I got close to taking my eye out so I gave up, but I always speculated at what wonders that geode might have contained (I think it got lost in one of the house moves).

Now I collect tools, because you never know when you might need one. I probably overcompensate because I have too many, so if you ever need a particular type of spanner, or an imperial Allen key, you know where to come.

*

As I mentioned already, I have always been a terrible insomniac. I wonder if this may have something to do with the dyslexia and possibly the ADHD I now have an official diagnosis for, because my mind is forever firing. I need distractions to calm it down, so I will watch something on television or listen to a podcast to help me relax, otherwise I am in danger of being overwhelmed by the constant chatter in my head. Sometimes these thoughts take a dark turn, and I dwell on things I may have perceived to be wrong, reliving and regretting my actions. This also keeps me awake.

It was harder to deal with the insomnia as a boarder because we were made to go to bed in our dormitories at a certain time every night, which was the moment when I missed my father the most. There were punishments if you did not adhere to the bedtime rules. We had overnight staff who were called 'nannies' and they were in charge of us. I found it torturous to get into bed at 7pm, especially in the summer months when it stayed light until 9pm. It reminded me of the Robert Louis Stevenson poem 'Bed in Summer', which I read then and can still recite now:

> *In winter I get up at night*
> *And dress by yellow candle-light.*
> *In summer quite the other way,*
> *I have to go to bed by day.*

During my first visit home, I explained this issue to my father, and he armed me with a torch with spare batteries for my return. It meant that after lights out I could go under the covers and attempt to read my books to take my mind off things. At a future school, I would deal with this by becoming a member of what we called 'the midnight gang,' running around the corridors while everyone slept and daring each other with challenges. For example, we might set a task like getting a bit of bread from the kitchen without getting caught. In an old school full of creaking floorboards, this was quite a feat. But at Ladymede my torch was a lifesaver, my father looking after me even when I was away. Incredibly, I never got caught. Probably because everyone else was asleep, including the nannies.

CHAPTER 2

There were not that many students at Ladymede – I think around 50 boarders – and what we saw as the flighty daygirls would come in and then run back home to their parents at night, unlike us brave boarders. There was one large dormitory for the younger students, which slept 12. The older pupils were put in smaller rooms with fewer people, which was a bit more exclusive – as was befitting their status. In the 12-bed dorm the beds were lined up, with a little bed under the window which seemed like an odd place for a bed, but luckily I never drew that short straw.

Saturday was laundry day, when we were given one freshly washed sheet and would have to remake our beds, taking the sheet from the top and putting it on the bottom of the bed; the bottom sheet would be sent to the wash and the clean sheet would be the new top sheet (this was a time before fitted sheets). It was quite an efficient system.

Saturday also meant a trip to the tuck shop, with our school cheque books, and we wrote cheques for the grand sum of five pence for a bag of sweets. We could also spend money at the Christmas fair or summer fete. The cheque books were to prepare us for the grown-up world, but I was never good at balancing them, and now cheques seem very dated.

I celebrated a couple of my birthdays at Ladymede, and one in particular stands out because I got in trouble before the day had even begun. My best friend, Louise, and I had gone for a wander to the outdoor classroom and missed the breakfast bell. When we realised, we raced over to the dining hall and skidded in as they were about to clear the plates. In my opinion it wasn't worth the effort because the tea tasted like dishwater – it came in a vat and was not strong and sweet as I liked it – and the porridge was gruel, but it kept us going until lunch. One of the teachers said we would miss breakfast as punishment.

'But miss,' Louise pleaded, 'it's Margaret's birthday today, she can't miss her breakfast.'

'I don't care if it's your birthday.' The teacher turned to me. 'You're late.'

Louise was so upset on my behalf, but I was thrilled, and she couldn't understand my reaction. I loved the fact that we were doing something a bit different on my birthday! This sort of weird optimism has bobbed along with me my entire life, and helped me turn difficult situations on their head.

*

I was facing more challenges in the classroom and assumptions were made about the Black kid from London who found it harder to learn. As we know, nobody mentioned dyslexia to me. But there were some wonderful teachers at the school who went above and beyond, giving me extra lessons and support, like Mrs Harding, who asked me to write a journal every day to help with my spelling and punctuation.

CHAPTER 2

I hated doing it; when one is in a confined space like boarding school, every day is pretty much the same, so it was hard to think up interesting things to write in my journal. I was given a small notebook, with wide lines and I had to write in pencil because I wasn't good enough to use a pen. Every day I scribbled down what I had been up to, similar to the letters home I had to write which were deathly dull and consisted of 'I did this' and 'I did that' sentences. My handwriting was neat, but my spelling was still atrocious. This is something that plagues me still. If I had terrible handwriting I could probably get away with bad spelling, but each letter is well formed, so my inability to spell is really obvious.

My father's voice would be in my head as I sat there, dejectedly bent over my work. He always said education was the key and he was paying a lot of money for me to be at the school, so I felt I was letting him down when I was not doing well. I don't know if I thought I was wasting his money, but I knew I should be trying harder. This was a contributing factor in being more helpful with matron, who oversaw all the nannies.

Up until this point, I had been one of those children who got out of bed at the very last moment every morning. It sort of made sense when you considered that I was up most nights with my torch reading. But I had a change of heart and I wanted to make a contribution. So now I was up with the lark, offering my services to matron along with the good students.

BOARDING SCHOOL BLUES

There was no ulterior motive for my sudden selfless behaviour, even though what I am about to say will make you think otherwise. We were split into four different houses – red, green, blue and yellow – and we could earn badges which equated to house points. There was a courtesy badge which I was keen on being awarded and then a half courtesy badge was introduced, so I achieved that before going full courtesy some time later. I promise this wasn't the reason for my sudden early starts, but I admit I wouldn't believe me if I were you.

A big revelation at Ladymede was my passion for acting. I adored it. I played three parts in a musical, *The Children's Crusade*, which involved a lot of singing and dancing. I can still remember some of the songs and will often sing them in my head or out loud, which is surprising considering how little I could retain educationally. One of my roles was a teacher who explained how you can measure the volume of pyramid. I had been provided with a model prop of a pyramid that had been specially made for the play. There was a lot of detail in my speech, and I kept getting it wrong. At a rehearsal, my maths teacher pulled me to one side and said the reason I was making mistakes was because I didn't understand what I was saying, so she explained it to me, and it made perfect sense. The lines just flowed then and I was full of confidence for the play. I did not need to regurgitate a complex script; I could just explain it in my own words because I knew what I was talking about.

CHAPTER 2

My favourite time in class was the round-robin storytelling, where each of us would contribute a line or two to make a story and at the end we were asked if we wanted to add to it, and my classmates would say, 'Can Margaret tell us some more?!' It was an ability I was proud of and could rely on, which gave me a worthy place among my peers. I was also the class clown. I might not have been good at the academics, but if I could make people laugh then the ice was broken and we could get on. It is something I still unconsciously do, in meetings, with strangers on TV and in difficult or challenging circumstances. If we can have a laugh then things rarely seem so bad. For my role as the teacher in the play, I won the annual drama award. My prize was to have my name inscribed on the drama shield for posterity, although what I really wished for was the book token the runner-up was awarded. I didn't get to see my name on the shield because after that term I left for another school: the voucher would have been much more useful. Although I do like to think of my name on that shield in a dark cupboard somewhere.

When I was ten, I had a careers chat with one of the teachers. I had spotted a book with an astronaut on the cover, and I thought, *This is me. This is what I want to be.* Here was a job that could allow me to literally reach for the stars. I mentioned this to a teacher, and she paused and then said, kindly, 'Why don't you think about going into nursing, because that's also science?' Remembering this reminds me of something Maya Angelou once said, although exactly where I can't recall: 'The worst thing you can say to me is "you can't do that", because then I know that that is exactly what I want to do.' This best describes my attitude then and I am still motivated by it today. You can underestimate me but I will succeed.

Quite soon after this conversation, I was in a science class. I remember it as if it was yesterday. The teacher told us that one litre of water weighs a kilogram and asked us to calculate the mass of one cubic centimetre of water. While others in the class were pondering the question, I got it immediately. I knew that a cubic centimetre of water is a thousandth of a litre. A thousandth of a kilogram is a gram so one cubic centimetre of water should weigh one gram. This felt easy so I stuck my hand up to answer, an unusual move for me from my position at the back of the class, but when I looked around the classroom, I was the only one with my hand up. If everyone was finding this difficult then I assumed I must be wrong, so I hastily put my hand down. Then I thought, *Oh, just give it a go*, and I got the question right. The teacher couldn't hide her surprise; I could barely hide mine. I think this was the moment when I realised I wasn't as stupid as everyone thought. In fact, I was the only pupil in the class to know the answer. That was a great feeling and one I wanted to repeat.

My father had always believed in me, but for the first time I now believed in myself. Somewhere tucked away

CHAPTER 2

was a confidence that would suddenly burst forth even if I wasn't outwardly displaying it. It's part of the story I tell people now. Tend to and grow your confidence because this is when things happen. I began to feel a little better about my abilities.

Oh, but my little sister, Gracie. She was desperately unhappy in Nigeria. Father would write to me telling me this and it tore my heart in two. I was powerless to do anything about it and couldn't bear the thought of her so far away from us all.

When I received another letter from my father, talking of Gracie's misery, I started reading it but felt so sad that I had to put the letter to one side. I went off to my lessons and tried not to let it preoccupy me throughout the day, planning to return to it when I felt strong enough to finish. When I picked it up again in the evening, full of dread, I read that my father was cancelling the arrangement with his relatives and bringing Gracie home. Hurrah! I was ecstatic and went running around the school. It was such a relief to hear this and a lesson to me in getting the whole picture before making assumptions.

*

I can count the time at Ladymede in two summers: the heatwave of 1976 which occurred soon after I arrived there and the Queen's Silver Jubilee in 1977. We were allowed a day off school and were taken to a street party in Thame, where we occasionally had school trips to my favourite children's bookshop, the Red House, which was based there.[1] It was a patriotic event, with red, blue and white decorations and a general feeling of community. It was the first time I was really aware of a monarchy and being part of a nation of people.

By this point, I was about nine years old, Sue and Hal were teenagers settled in their boarding schools and four-year-old Gracie had returned home to England, but not to live with our father for long. He didn't send her to the local London school. Instead, she went to live with two elderly sisters, not far from me in Chinnor, Oxfordshire, and attended the village school there. I never questioned it at the time but in hindsight, this seems like an odd decision. Father continued to work full time and beyond, doing a number of jobs, so placing Gracie near me probably made sense. I have no idea how he found the sisters, but I was

---

1. I recently attended a party celebrating 90 years of Penguin Books with the editor of this book. It was a wonderful party with hundreds of people in attendance, and I felt very privileged to be there with the great and the good. They had a curated exhibition of the history of Penguin and I got talking to a wonderful archivist who knew about the shop in Thame and reminded me of the Puffin Club, a book club for young people of which I was a member.

CHAPTER 2

thrilled because I could visit Gracie on the occasional weekend or stay for a couple of nights in the school holidays.

Gracie wasn't the only lodger who 'Auntie' Olive and 'Auntie' Mary took in. There was another Black girl, Toyin, who was older than us both, and then the housekeeper, Meg. The huge, airy house was called Greenacres. I had no idea what an acre was, but if you walked through their garden there was an archway in one of the bushes which led to a much bigger expanse of green. Even now when I think of an acre the image of that green field comes to mind.

It was an unusual set-up, Gracie being in this house of strangers, but they were kind, and I was welcomed when I stayed. She was overjoyed not to be in Nigeria, so she wasn't complaining.

Olive and Mary had a neighbour friend who would pop over for a cup of tea, bringing her little dog, Yip, for Gracie to play with. Once, when I was staying, Meg the housekeeper had a bone for Yip and asked us to take it to the neighbour's house a few doors down. Gracie knew where we were going so we set off on our quest, the bone wrapped in plastic in my hand. We wandered down the street, and we kept on walking and every so often I would ask Gracie, 'Is it this house? Or this one?' Each time she answered no, so we kept on going until we were maybe half a mile away. At this point, we turned back and eventually reached the house, which was probably next door but one to Greenacres. We delivered the bone but when we got back to the house, the police were there. Whatever could have happened?

Apparently, the police had been called because two kids had gone missing. Oh! They meant us. Olive and Mary were so worried they had reported us missing after checking with their neighbour and discovering that we had not made it to her. We didn't get into trouble; I think everyone was just relieved we were back – and, most importantly, Yip got his bone.

During this time, Father had set up his own import–export business, buying shoes in Italy and selling them to Nigeria and the UK. I think this was going quite well, and at one point I think he even made a deal with Harrods to sell the shoes. This was a bit of a side hustle though as he hadn't given up his job at Pizza Express, where he was now a manager. He would bring samples of the shoes home and most of them were too big for my little feet, but there was one pair of high heels that fitted. I shouldn't have been wearing them, but I couldn't resist and would walk around pretending to be a sophisticated grown-up.

There must have been some money at that point because Father bought a brand-new house in Aylesbury, pretty close to my school. He took me there twice, just him and me. He seemed excited as we walked around the empty rooms talking about how we would decorate and who would have which bedroom. It was thrilling because I thought we could

CHAPTER 2

be happy together in a place like that. There was a dead red admiral butterfly trapped within the double glazing. I couldn't work out how it had got there, and I couldn't take my eyes off this beautiful colourful insect, imprisoned between the two sheets of glass. I have never forgotten that. Unfortunately, we never did move into that house. I think he must have had to sell it soon after.

*

Things began to change for me at Ladymede after a couple of years. When activities were being organised, teachers would suggest that I may not want to take part. They would gently tell me I couldn't afford it. We might have been planning a trip to the Red House bookshop or renewing the Puffin Club membership and a teacher would say, 'Perhaps you shouldn't buy a book today, Margaret.'

Not only was my father struggling to pay for additional costs, but he was also beginning to find the fees a stretch. It took a while for this to sink in, but when it did, I understood, and I was happy to step out of extracurricular events. Nobody ever made a big deal of it, but I began to feel like I was outstaying my welcome. For the first time since I had been at the school, I was hit by imposter syndrome, something that has reared its manipulative head on occasion throughout my life. I realised I didn't really fit in to this type of environment, and change was on the horizon.

I loved stationery; I still do. Whenever our teachers allowed us out to go to the shops, I would head straight to the stationer's to buy a new pen or a notebook. One day, I was in a classroom which was randomly tucked in one of the school outbuildings and I spotted a pencil case on the floor. I picked it up and had a rifle through, marvelling at the lovely contents. I kept it. I should have handed it in, but I didn't. I took it and then I got caught. It was awful and I was reprimanded, but not punished. In fact, lovely Mrs Harding bought me a small pencil case and filled it with pens, pencils and erasers. It was a beautifully kind gesture and I wish that the story had ended there.

Quite soon after, my friend Louise discovered some of her family photos were missing; they had been in a small plastic frame by her bed and they were presumed stolen rather than lost. Unfortunately, I was accused of taking them because I had what could be described as 'previous form', and worse, I had found part of one of the transparent plastic photo covers on the floor of the loo and – magpie that I am – kept it. I know this must have looked bad. If I were a teacher at the school I may have assumed the worst, but I didn't take the photos. The teachers wanted me to confess, but I couldn't hold my hands up for something I hadn't done. Luckily Louise believed me – she knew

CHAPTER 2

I hadn't taken them – but it turned out we didn't have much more time together.

*

That year I came home for the summer holiday, and I never returned to Ladymede. I think this was because of a lack of money, not due to me being asked to leave. There was no opportunity to say goodbye to my friends and, when I returned home, it wasn't to the house we had lived in before, it was to a bedsit in Hampstead.

This was to happen quite often, and generally without any warning. Most of my belongings would be packed up in my absence and transported to the next place, but often things were left behind or thrown away, so I had very few toys or belongings in general. The bedsit was a tight squeeze for five of us anyway, so it made sense to leave stuff behind. I do not need a psychotherapist to tell me that this may be why I am now such a tenacious hoarder. I hang on to so many things. The objects feel like my tenuous past. I am frightened they will disappear, and I keep them for sentimental reasons or because they trigger memories for me. Holding tangible objects and surrounding myself with cherished items is a response to a peripatetic childhood, where I had very little and couldn't rely on my things being there in the future.

BOARDING SCHOOL BLUES

It has impacted my own parenting in a good way, I hope. I am forever discussing things with Lori and explaining situations. I think it drives her mad, but I don't want to be the mother who says, 'Because I say so' and makes decisions without talking to her. Maybe I go too far the other way and spend too long sharing my thoughts and opinions. As a child, I was always underestimated, and I do not want her to feel how I did. The assumption was that I didn't know what was going on around me, and I remember how frustrated I was by this. It has made me think deeply about how I communicate with my daughter and with children at events. I think it is one of the reasons I am good at speaking to kids: I talk to them as equals. They know so much more than we realise and they deserve our honesty and respect.

The Hampstead flat was just one room, with a large bed where Father, Gracie and I slept, and a sofa bed which Sue and Hal slept on. There was a shared bathroom out in the hallway, but I'm not sure how many other residents used it. When my cousin, Bumi, came to stay for a while it seemed like a crazy idea because we were already a crowd, but her visit coincided with one of the upstairs flats being vacant because the owner was away, so we were allowed to use his bedsit in his absence.

The bedsit was above the Pizza Express, which my father was now managing. Sometimes, we popped down for dinner, revelling in our familiarity with the staff and the fact that home was just upstairs. Everybody knew us there.

CHAPTER 2

Gracie and I sat at a tiny table for two, close to the till, and waited for one of the front-of-house staff to bring us pizza. We didn't need to order, and we weren't asked to pay, but we would leave a penny as a joke, saying, 'Keep the change!'

Being crammed in a tiny living space meant we were outside a lot. What we missed in a bigger, cosier flat we more than made up for with Hampstead Heath as our garden and we spent a huge amount of time there, staying out all day in the summer and eating ice cream. There was a small permanent fun fair where we would spend hours on the bumper cars if nobody was around, and a lake where people rented boats and rowed out. It was one of the rare times when us four sisters were together, united in a common goal of halcyon summer fun. We had a blast with Sue and Hal in charge of plans and the budget while Gracie and I tagged along, trying to keep up.

Even now, when someone asks me where home is, I automatically think of Hampstead Heath, which was the happiest backdrop to so much of my childhood, no matter which house we lived in. This is where Sue introduced us to the music of Stevie Wonder. We would dance around or just listen to the powerful lyrics. Even today I am amazed at the feeling of carefree fun his songs evoke.

*

BOARDING SCHOOL BLUES

At the beginning of the autumn term, Sue and Hal went back to their boarding schools, Gracie returned to the sisters at Greenacres, and I went to a new boarding school. Moyles Court was in Ringwood, Hampshire and was a mixed school, which I think was unusual at the time. It was very different to Ladymede, a bit more down to earth and probably a little cheaper. I arrived as a childlike ten-year-old and left eighteen months later in the early throes of puberty.

At Moyles Court, I spent a lot of time in a state of angst, worried about how complicated the world was and crying over the smallest triggers. It wasn't particularly dramatic, just very confusing and adolescent. One of the saving graces there was my science teacher, Colonel Andrews. He was quite a character, strolling around the corridors with a pipe, and he had one of those rude pens in his top pocket, which revealed a scantily clad woman if you turned it upside down. It would be considered very inappropriate now but at the time it made us laugh. He took us out sailing in tiny dinghies at a local lake and was in charge of the swimming lessons. He told me he had high hopes for my dives. I would go up to the diving board, stand correctly and make all the right moves as if about to perform the perfect swan dive, before belly-flopping into the water. He ribbed me for that, laughing about my brilliant set-up and terrible execution!

Best of all was how he would encourage us to do practical, hands-on science, like dissecting a frog, collecting

CHAPTER 2

marsh gas or identifying plants on walks. He took us to a large pond in the school grounds and showed us how to submerge an empty jam jar into the water, let it fill completely and then invert it. Then he poked the boggy silt at the bottom of the pond and collected the escaping gas that came out, before setting it on fire. We were all mesmerised – me more than most. In fact, once I was too close to the edge of the pond, got overexcited, and my foot slipped, resulting in one light summer sandal with ankle sock sunk into the slime. It stank but I didn't care. I hobbled back to class with my mind firing with possibility.

I have never forgotten Colonel Andrew's practical attitude to science. I am a huge advocate for kids learning through hands-on experiments, and it is a shame that there is much less of it about these days. If we take these mind-blowing moments out of the curriculum it makes generating the scientists of the future a lot more challenging. Very few students can get excited about physics, chemistry or biology if they are confined to reading about the experiments in textbooks or just watching videos.

There were three things around this time that influenced my destiny. The first was the realisation that it was scientists who put people into space. The second was the people around me – my father, the Colonel and other teachers – who inspired and encouraged my passion for science. Thirdly, being dyslexic made me curious and logical, boosting my appetite for practical experiments and 3D awareness.

BOARDING SCHOOL BLUES

I think the combination of these factors enabled me to pursue and realise my crazy dream of studying the universe, and gave me a better aptitude for the subject.

There is currently some research being done by Dr Helen Taylor at the University of Cambridge, which suggests that people with dyslexia often have traits that make them well-suited to explore the unknown. It also postulates that this is likely to play a fundamental role in human adaptation to changing environments. When I met Helen recently she told me about her research, and I am looking forward to exploring this further.

*

I lived in a maisonette in Southgate with my father and sisters for a brief time. I think I went to school there too. Father didn't always have enough funds to pay the electricity bill, so when this was the case, we used camping gas lamps instead. Just like my father to find a practical solution to a problem. They had small mesh wicks about the size of a mushroom which glowed, giving off a comforting warm yellow light, and heat too. We even cooked on a camping stove in the kitchen. (Many of my contemporaries speak about camping holidays with their parents. I like to point out that I did this too, but without the vagaries of the outdoors to spoil the fun.) The whole thing felt like a bit of an adventure, as

CHAPTER 2

if we really were camping, yet I don't remember ever being cold or hungry. It wasn't that sort of challenge.

We then returned to the bedsit in Hampstead. My sisters were at their various schools, but for the first time ever, I had no allocated school to go to, so I stayed home with Father.

Looking back, I am not sure why I seemed to be the only child to be moved around all the time, compared to my other sisters. I was the only one who really ramped up the school count. Suddenly, I had lucked out because I had my father and our bedsit all to myself. That was until the truancy officer started asking questions about why I wasn't in school. It was a shame, because I loved being at home with my father.

Around this time, Sue must have been home for half term or something, and she decided to get me out of the house so she took me out to the cinema. The film I really wanted to see was *Alien*, the tag line 'In space no one can hear you scream' really caught my imagination, but it had a certificate of X (18 equivalent now), so I knew that it was not going to happen. As we stood at the box office Sue asked for an adult and a child ticket. The guy just stared at her as if she was mad. Sue had clearly not realised the age rating of the film. I think we got to see something more 'appropriate' but I love that she gave it a go. When I saw the film many years later, I knew it had been well worth the wait.

The holiday could not last forever, and eventually Father found me a place at a school and a family to stay with.

BOARDING SCHOOL BLUES

Looking back, I guess it was a weird solution, but considering the alternatives, it wasn't a bad one. I had feared I would be sent to my aunt's house to live with her and my cousins, so living with kind strangers seemed like a nicer solution.

By this time, I was in secondary education and about to start at another new school when I was parachuted in to live with Yvonne and her children – a girl of similar age to me and her younger brother. I assume Father must have paid my bed and board. I have no idea how he found them or arranged the foster, nor do I remember meeting them before I moved in. It may have been a link through the church he attended. Thankfully they were a nice family. I was particularly fond of their goat, Snivage, who lived in their back garden and enjoyed eating scraps. She was quite young, very sweet and used to bleat loudly if we went out and left her in the garden. Her bleating sounded like a small child crying, so I can't imagine what the neighbours thought.

I went to a school near Yvonne's house and started again, adapting and changing to my new norm. Making friends and walking back to a place that was not really my home, to live with people I didn't initially know. I didn't question this set-up, but I felt like an adjunct to the small family. They never made me feel like that, but it was discombobulating. I had been inserted into other people's lives, even though I was aware that this was a necessity. I can't remember any resentment or anger at the situation, just the familiar feeling of change.

CHAPTER 2

A while ago, one of Yvonne's daughters made contact with me and I had every intention of replying, but when I came to do so I found I couldn't. I find it very hard to re-acquaint myself with even my recent past, let alone people from so long ago. Just like the Doctor in *Doctor Who*, I feel that I have been through many regenerations since this time, and the person that they knew no longer exists. What if she was an adaptation, designed to blend in with the circumstances? Maybe this does not matter, as everyone changes through experience and time.

Also, I wasn't sure I wanted to revisit this time in my life and be awash with the emotions of everything that was going on in my family during that period. Just like my role model Spock from *Star Trek*, I was quite good at supressing my feelings, but a backwards glance now risks releasing a torrent of subconscious emotion.

After about six months, I returned home, leaving Yvonne's family behind.

Chapter 3
Dunce Dunce Double D

It probably seemed a bit strange that a Black girl from inner-city London was interested in the subject of space. As a child and for most of my adult life, I thought astronomy was done by white guys in togas. When I was learning about the history of astronomy, I mainly heard about the Ancient Greeks and Romans – a very Eurocentric view of the history of the subject. But more recently I have discovered a whole subject area called archeoastronomy, literally the archaeology of astronomy. It is a subject that I find fascinating. I sometimes describe it as 'who knew what when'; I think it would make a great documentary. It is amazing how every culture has looked up at the night sky and wondered what was going on out there. Virtually every culture has creation stories linked to what can be seen in nature, and especially the night sky. It is truly the heritage of us all.

I try to relay this understanding as part of my public-speaking assignments to audiences young and old. I like to create an image in my mind's eye to share with everyone. The picture is of ancient cave people, sitting around

CHAPTER 3

a campfire at night, grunting at each other because formal language isn't a thing yet, but what unites them is looking up at the beautiful night sky. Maybe like one of the earliest streaming services for our ancestors around the world. Through the mystery of the cosmos, early myths and legends form as people endeavour to understand what is going on out there.

Then I move on to Aristotle, with his belief that the earth was the centre of everything. This made sense to people at the time because you can see the sun rise and set – the sun seemed to go around the earth. But it wasn't just the sun; if you could find the pole in either hemisphere, then all the stars seemed to wheel around that, except for a few that seemed to move forwards in space and then backwards over a period of months. The ancients dubbed these the wandering stars.

As the years whizzed past, these 'wandering stars' became of great interest to people like the Renaissance astronomer Copernicus. Using his own and other astronomers' detailed measurements, he postulated the theory of the sun-centred (or heliocentric) universe and concluded that it was the sun, rather than the earth, that was the centre of everything. He suggested that the wandering stars were planets which, like the earth, travelled around the sun.

By this stage in history, along comes the telescope. Many people think Galileo invented the telescope. He did not, a German–Dutch spectacle maker called Hans Lipperhey was the first person to apply for a patent for his design of the 'looker', an instrument that used two lenses enclosed in a housing. Although he was paid by the local government for the design, the patent was not granted as it was thought that the ideas behind such a simple system would be hard to protect. Galileo was one of the first people to use this instrument and observe the night sky. He could see the rings of Saturn and the larger moons moving around Jupiter, which were named the Galilean moons after him.

The telescope was a great improvement on the naked eye. It meant that people were able to gain a better understanding of the cosmos by recording more detailed data, and were able to log and catalogue their findings. The Danish astronomer Tycho Brahe made incredibly detailed calculations by observing the passage of astronomical bodies as they moved across the night sky. He was known for having a fake nose of silver because he had lost his real nose in a sword fight. Luckily, he had the money to afford the luxurious replacement, as well as cutting-edge astronomical equipment. Before he died, he worked with the German astronomer Johannes Kepler, and they discovered that the orbits around the sun were not circular but elliptical, like a squished circle. This was quite a controversial revelation at the time because the circular orbit was thought to be a thing of God.

Isaac Newton was up next. He was quite a character and a game changer in the story of our understanding of the solar system, in so many different ways. He realised that

CHAPTER 3

white light from the sun is made up of a rainbow of coloured light through a simple experiment of putting a hole in one of the shutters at his bedroom window, creating a shaft of sunlight and then using a prism to split the light into its component colours. Then he used another prism to recombine the light to turn it back to white.

Added to these great discoveries, he transformed the telescope. The telescope that Galileo used had lenses to magnify the objects under observation. But in those days even the best glass available wasn't of good quality. Due to the manufacturing process, the glass contained inclusions and air bubbles which would cause aberrations in any image produced. Newton came up with a new telescope design that would also use glass, but as a reflector instead of a transmitter of light. With this design, it was the shape of the mirror that provided the magnification and the light no longer passed through the glass. Modern telescopes – whether amateur or professional, ground or space based – still use the Newtonian design and its variations.

Arguably, Newton is best known for coming up with the theory of gravity, which is one of the tenets of science. (I have been lucky enough to sit under the very apple tree where he made his great discovery.) I consider him a kindred spirit, and he has featured heavily throughout my life. A Lincolnshire lad whose father died before he was born, he was brought up by his mother, who wanted him to be a farmer. There are letters written by her that show her despair at his ability to look after the sheep who would regularly escape because Newton had his head in the clouds. Incredibly he wasn't considered very bright at school, and yet he went on to do great things.

It is fascinating to look back for hundreds and then thousands of years, thinking that every culture across the world has looked up at the stars and wondered. We can connect with this magic if we go out on a clear night and stare at the cosmos around us.

A few years ago, I was doing some work with English Heritage and I visited Stonehenge, which is thought to have been an astronomical observatory. I met the woman who had the enviable title of The Keeper of the Stones, as I was working on a project to film the night sky there. At the end of the shoot, she asked me if I would like to stand in Stonehenge on my own. I jumped at the chance, so she and the film crew withdrew, and I moved into the middle of the circle. There I was among the stones, immersed in the silence, majesty and spirituality of this incredible place, and taking it all in so I could hang on to the feeling of it for ever.

A couple of weeks later, I returned with Martin and Lori for the Summer Solstice. It involved a very early start for us – needing to get up at around 3.30am – which was a hefty ask for eight-year-old Lori, but she was game, caught up in the excitement of the adventure. We were rewarded. It was amazing to see the dawn sunlight on the longest day of the year pass through the key stones, just as it was designed to do.

CHAPTER 3

For this visit, I shared Stonehenge with around 10,000 other people, but it was just as magical as when I was there alone, albeit in a different way.

Many people think Stonehenge, at around 5,000 years old, is the oldest stone circle in the world, but that title is disputed. Some believe that Nabta Playa, which sits on African soil and is over 9,000 years old, is the earliest example of a stone circle used for astronomy found to date. Stone circles, although perhaps the most well-known, are not the only ancient astronomical tool. In the ancient world, there were these wonderful instruments called astrolabes, which were a way of tracing the movement of stars and planets with great accuracy. Originally developed by the Greeks, they were later perfected by Arabic astronomers in the Islamic Golden Age to determine prayer times and the direction of Mecca. Not only was it a clever piece of scientific equipment, it was often also a piece of art with beautiful engravings. I call it STEAM – science, technology, engineering, art and maths all rolled in together – and around the rim of many astrolabes, there are words written in Arabic and from the Koran. It is a celebration of a diverse culture, all in one instrument.

It's a thrill to see people's eyes light up with new knowledge and understanding when I cover these areas in my talks. It's why I do what I do. Like the best type of lesson. Now, class dismissed!

*

DUNCE DUNCE DOUBLE D

So how did this Black kid in London get so fascinated by the universe and its history? Amazing icons helped, like Yuri Gagarin, the first person in space, and Neil Armstrong taking one small step for man and one giant leap for mankind as he stepped out onto the moon's surface. I heard about these people and what they achieved, and I thought, *Me too! I want to experience what they did.* This unquenchable desire to go into space has been the driving force of my life.

From a very young age, I would look up to the sky and visualise myself zooming through the clouds, then up above Earth's atmosphere and out into the sparkling cosmos. I imagined swirling around in the darkness, looking back at the little planet I had left behind and marvelling at such a beautiful place full of incredible landscapes and amazing people. How lucky we are to live on Earth! And then I would head to the moon, my ultimate destination, and my dream would come true.

As I have mentioned, my father was fascinated by the universe and particularly loved the moon. He lit the touchpaper for my enduring lunar passion, which was further fuelled by *The Clangers*, and then Sue introduced me to *Star Trek*. That programme underlined another reason why the cosmos captivated me: race doesn't matter there. When you're in space you see the world as a globe, a whole perfect thing, without country borders, barriers or boundaries. This wasn't my experience of living on it. Growing up, I didn't really feel that I belonged anywhere, caught between

CHAPTER 3

Nigeria and the UK, and yet *Star Trek* showed people from diverse backgrounds working together towards a common goal. It made a big impression on me. As did Lieutenant Uhura. She was one of my role models, and I was speechless when I met Nichelle Nichols, the actress who played her in the original series, many years later.

Space travel also represented the idea of freedom and escape. I could remove myself from the trials and tribulations of my home life and soar high above custody battles, older-sister control and education issues. I wanted to shake off the tumultuous and challenging days that bound me and be transported to another realm.

This may make it sound like I had a traumatic childhood. I did not. There was happiness among the dysfunctionalities. I knew I was loved. I always had a roof over my head and enough food to eat. Yes, it was a complicated upbringing, but I know of far worse out there. Also, as a child you have no other reference point: you assume your experience is the norm, and if you do compare yourself to your contemporaries then it is their life that seems strange, not yours. When you are in it, you don't therapise it. You find a way to survive and, hopefully, thrive – and mine was to daydream my way around outer space.

*

DUNCE DUNCE DOUBLE D

From the age of seven to twelve, the number of times I visited my mum can be counted on one hand. Then there was a flurry of court activity and another big conversation about who Gracie and I wanted to live with. Mum was now settled in Hastings with Barry in his new parish. Father was working hard to support us and, as a result, was home less and less. It was wonderful when he was home, but then he was busy looking after us. He was exhausted and it was taking a toll on his health.

As we approached another custody hearing, Sue and Hal took me to one side and pointed this out to me, explaining that although it was good for us all to be together, it probably wasn't good for our father and perhaps the pressure of looking after all four of us was taking its toll. Again, it made sense, but it broke my heart to leave Father. It also came with that piercing guilt. I didn't want to hurt him by leaving, but staying would hurt him too.

I reasoned that I was making the decision for the good of my father and also, as it had been suggested (with good reason) by Sue and Hal, I could delegate responsibility and ease the guilt I felt at leaving him.

For Gracie and me this meant a plunge back into our mum's world. Mum and Barry must have found it strange to suddenly be parents to two growing girls, one edging into adolescence. I wonder how our arrival was explained to the parishioners, and whether they knew we were Mum's daughters or if they assumed we had been adopted?

CHAPTER 3

I have no idea who they thought we were, but we were never made to feel uncomfortable in the community.

I refused to call Barry 'Father', even though he and Mum wanted me to. I already had a father, so that name and position was reserved for one man and one man only. It was doubly odd because Barry was called 'Father' by his entire flock. I just called him Barry.

The vicarage was perched on a hill a couple of miles outside of Hastings. It was a nice place to live, like something out of an Enid Blyton book, with geese in the garden who would chase the parishioners as they arrived for church next door. We went down to the sea at the weekends. During the summer it was quite touristy there, and Barry bought us ice cream and candy floss. In the winter, it was stormy and empty of everyone but the locals, and I loved that. I would stand under the defunct pier lashed by the spray coming off the sea. Nature at her most raw. I loved the variety of the weather in such a place.

Father was never far from my mind, and I would try to find people who looked like him in my everyday life. This was the 1980s and there was not much Black representation on TV, so I used to look to the chap on the Uncle Ben's packaging. I also considered Nelson Mandela a father substitute. I'm not sure if Father ever understood why we left, but leaving did create a large emotional hole in my life.

Also, having been brought up with a fear of my mother, living in the vicarage was fraught with anxiety, especially at first. Gracie and I had separate rooms for the first time in our lives, but would sneak in together as it felt there was likely to be safety in numbers. As time went on with no supernatural occurrences we relaxed into a new way of life. Even so, my eczema and asthma flared up due to the stress of it all.

The time I spent in Hastings could be considered something closer to a more conventional childhood. By this age I was old enough to go into town to meet friends, and we would often go to the cinema, clutching packets of sweets and purchasing big bags of popcorn. I learned to ride a bike in the back garden of the vicarage. It had a slope which I could use to pick up speed, then get my feet on the pedals and cycle for a bit before slowing down and getting too wobbly to continue.

The church itself was a large red-brick building and also my playground. With permission. I would head inside and play the organ badly, trying desperately to remember the piano lessons that I'd had at various schools. The noise was amazing, a wall of sound reverberating off the walls, even if it was only 'Chopsticks'. Of course, anything played on an organ sounds much more impressive, so even the simple nursery rhymes that I picked out took on a gothic grandeur. Barry practised what is called 'high church' in his services, leaning in to the pomp and ceremony. It involved lots of candles, robes for himself and the altar boys and one of my favourite things, incense. Just the smell of it today evokes

CHAPTER 3

memories, not just of the church in Hastings, but of my father and the altar we used to pray at in later years.[2]

Mum knew that my father valued education highly, and that we had been in private schools before we left his care. She wanted to keep us in the manner we had become accustomed, so arranged for us to attend a nearby girls' school. Although my mum worked, it must have been a stretch for her and Barry to have two children in private education.

Mum would drop us at school every day, in her Citroën 2CV, which we referred to as the upside-down pram on wheels. It was like being driven in a sewing machine. We were at a Catholic Convent School, St Mary's on the Ridgeway, which took day students as well as boarders.

As it was a convent school, there were a lot of nuns. We broke them down into two categories. There were our teachers, who wore a sort of uniform of a 1970s skirt and jumper, and then there were the nuns in full habit who would trundle around the school. I believe these nuns had taken a vow of silence as we never saw them speak. We called them the

---

2. When my father passed away, one of the ways I would remember him was by lighting a candle in any of the churches I found on my travels around the world. Tucked away securely, I still have one of the blankets from his flat that smells faintly of incense. I don't get it out often as I am trying to conserve the aroma, but when I do need its comfort I am instantaneously transported back to him and my childhood.

clockwork nuns because it was as if someone had turned a key and set them moving. You couldn't see their feet.

I wasn't Catholic, but I was comfortable in a religious environment, so I learned the catechisms and had a set of rosary beads. Unlike Gracie, who had a rough time at school. She was teased a lot, and I used to come up with solutions to help her protect herself. Once I told her to tell them she was the pop star Michael Jackson's cousin, so she did. The girls came up to me asking if it was true and I said, cool as a cucumber, 'Yeah, didn't you know? We visit him occasionally.' I don't know how I thought we could pull it off, but it did the trick and they left her alone.

The school closed down while we were there, and we were transferred to a different one. Several of the thirteen schools I went to have now shut. It wasn't anything to do with me – at least I hope not!

During this time, nothing changed for me educationally and I didn't make good progress. I felt like I was failing the education system, and the education system was failing me. My father had this expression, 'dunce, dunce, double D', which came with the image of a child sitting in a corner with a big D for 'dunce' on their hat, and that's how I described myself. Nobody expected anything from me academically, and I didn't really expect anything from myself. Occasionally there would be a spark of joy. I entered an art competition run by the BBC children's show *Blue Peter*. I drew a spider which really worked as a pencil drawing and

CHAPTER 3

then unfortunately ruined it when I coloured it in, so the art teacher suggested I draw it again. I didn't think I would be able to do it second time around – it felt like a fluke creating something so great the first time – but I did it again and it was just as good. I am not sure if I was shortlisted in the competition, but it did give me a confidence in the subject. (In this respect I was more following in my mum's footsteps, as in later years she became a successful artist, taking commissions and creating original pieces.)

Although I did not get much opportunity, I still adored performing, so Mum found a lovely young woman who taught speech and drama locally. Miss Dyer was a wonder, and I loved going to see her; I found the sessions so energising. Finally, here was something I was good at. I fancied myself as a bit of a thespian at the time, although I would have been very surprised to learn that I would later appear on television and radio. My mother gave me a gift with these lessons that has proven invaluable in my career.[3]

I have always been an avid film buff. I loved films because they gave me access to stories without the slog of reading.

---

3. It is also something I have in common with my big sister Sue, who went on to feature in films and television dramas. Although she is an amazing actress, when I see her on screen I can't help seeing my big sister. She was once playing a baddie in a cop show and she had a gun. The police were after her and I got very emotional. I started to shout at the TV, 'Drop the gun, Sue! They're after you!'

DUNCE DUNCE DOUBLE D

Despite this I also loved books, and still do. For a couple of hours, I could be whisked away to somewhere different or imagine myself as a character in a movie.

Every year in Hastings, speech and drama competitions were held at the White Rock Pavilion, and I took part in a few different categories including prose and poetry. It took me a long time to learn the poems off by heart, but once I had they were chiselled into my brain for ever. I can still remember many of them and often recite them to Lori, whether she wants me to or not.

Those speech and drama lessons are still paying off. A few years ago, I was asked to give a reading at St Paul's Cathedral at Christmas. I couldn't believe it, to be in that space, full to the brim with people and reading an excerpt from the Bible. Last year I was asked to read a wonderful poem called 'Amazing Peace' by one of my all-time role models, the awe-inspiring Maya Angelou. As a dyslexic, reading out loud is not one of my strengths. To increase the pressure, I was giving my reading alongside acting royalty Dominic West, and the ultimate narrator and actor, Stephen Fry. I ended up sitting next to Stephen in the front pew and we had a discussion as to how water and vodka were very much related. When it was my turn to go up, I was very nervous and convinced that I would fall over as I climbed up to the pulpit. But once I was up there the old training kicked in, and I leaned into the drama of the piece. On returning to my seat, I was still quite shaky but pleased with my

CHAPTER 3

performance. I am not sure because of my slightly wobbly condition, but I think that Stephen said 'superlative' to me – which I hope was related to my performance.

We were happy with Mum in the vicarage, but it still felt like living with strangers. Barry and I had only met a few times just for an hour or two on our rare day visits, but Mum was a stranger too, mainly due to the rather binary way the custody had worked out. I was about four or five the last time I had lived with Mum, so moving in with her some ten years later felt odd. The chameleon in me did not know who to be.

Barry proved a good father figure to Gracie and me. He was kind, stable and interested in us, or at least, that was my experience of him. However, within two years of us moving in with them, their marriage was on the rocks. I don't think our presence had anything to do with this, but we were caught up in the fallout again. It had gone from being a harmonious house to an angry environment, and we were unsurprised when they broke up. Things were about to change again.

*

The second time I ran away from home, I was fourteen and far more cognisant of what I was doing compared with the first time.

DUNCE DUNCE DOUBLE D

While we were with Mum, we hardly saw our father. Sue and Hal may have come to visit once or twice, but as always our family was polarised. If we were with one parent, then we couldn't be with the other, that was just the way it was. I didn't want to make a fuss with Mum about seeing my father because I held residual guilt for the years I had been estranged from her, but I was incredibly uncomfortable about what her divorce from Barry would mean for Gracie and me.

Mum decided she was going to move away from Hastings and settle in a tiny Sussex village. Ultimately this was the right thing for her because she has been there ever since, but it did not feel like the right thing for us. The opposite, in fact. She was talking about sending Gracie to a school for 'special' educational needs, because she was behind in her learning too, and I didn't agree with the decision but was unsure how to argue against it. It triggered something in me. Not only did I think Gracie would not thrive there, but there was also a lot of uncertainty as to which school *I* would be going to. And as I was starting my GCSEs soon, this was a worry.

I was now a teenager who had grown up in the heart of London, so Hastings felt like a backwater to me. It had many charms, but I yearned for the hustle and bustle of city life that I had known when I lived with Father. Now we were going to be propelled into the middle of nowhere, with muddy fields and one bus a week.

CHAPTER 3

My father was living in a council flat in Belsize Park at this point, and was keen to have us back. Weighing everything up, I knew we had to return to him, one way or another. So for the second time in my young life, my sister and I were running away. I got us to the train station. We were meant to be in school, so we were in our uniforms. I think that was the one and only time I was responsible for playing truant during my schooling.

Once again, we were leaving a version of our lives and all our belongings behind. Father and I must have conversed beforehand because he knew exactly which train we were arriving on, and had set the wheels in motion for a new school for us both. He may have even sent us money for the train fare, but I think it was more likely that we saved up our pocket money. I remember looking out of the train window as towns and fields rushed past, Gracie sitting next to me, and knowing I was doing something that would again hurt a person I dearly loved. Father was waiting for us when we arrived in London in his old red Vauxhall Viva.

Back in Hastings, Mum was about to find a letter from me, explaining why we had left. I couldn't contemplate her reaction. Years later, she found the notelets I had used to write my departure letter – I think it had a cute little cartoon dog on the top corner – and she referred to it as 'that treacherous stationery'. She associated it with betrayal.

Leaving my mum behind was – and still is – a cause of sadness for me; we were close, so I know it caused her a great deal of pain, especially because she was already vulnerable. It could have been done in a better way perhaps, and although it was many years ago, it is something that will always be part of our relationship. It's a particularly hard part of the past to face.

I was a child, eternally stuck between two warring parents who I was regularly forced to choose between, and when I chose one it meant being estranged from the other. What happened to them as a couple and us as a family was not my fault. They were responsible for their own behaviour. All these things I know.

Both times I ran away, I was given an opportunity – a branch to grab hold of – but I was still a child, and I was not the originator of either plan, even though I could see how the escape could benefit Gracie and me. I give kids much more credit than many people do, but I was put in a difficult situation, and it's hard to know how I could have handled it differently.

The raw truth is that, if I had my time again, if someone were to turn the clock back to when Gracie and I ran to Hastings station, we would still catch the London train.

I have talked about my sensitivity and inclination to empathy overload, and strangely I don't think that this is in conflict with these feelings. I did and still do empathise and feel the pain of others, especially if I feel that I've caused it, but over the years I have learned to compartmentalise my feelings. It enables me to feel the emotions but contain them

CHAPTER 3

and do what I fundamentally hope is right, knowing that I will need to face the repercussions when they occur.

My early experience was in the school of hard knocks and this has helped shape the person I am, as our upbringing does for us all. It enabled me to bolster my determination, and it gave me the power I needed to survive in a world where changes and challenges were coming thick and fast.

I left Hastings in a hurry and never got a chance to return to the vicarage. Recently, I saw that the church was up for sale. For one mad minute I thought about buying it and turning it into a science centre. Now wouldn't that be amazing?! I decided not to add that to my overloaded to-do list, but it was appealing. I occasionally take Lori to Hastings and, as we drive around, I point out old haunts, so I do think fondly of the time I spent there.

*

The school Father found for us continued the Catholic theme: La Sainte Union Convent School, or LSU, where the teachers assumed we were Catholic, and we never said we weren't. It also meant that our daily commute was across my beloved Hampstead Heath. Welcome home.

At LSU, the teachers asked me what stream I should be in. They had not received any paperwork from my old school cataloguing my woeful ability, so they looked at me with fresh eyes, and I felt reenergised by the lack of preconceptions. I could reinvent myself again and switch from Maggie the Class Clown to Maggie the Grade-A Student. All my life I had been in the lower stream and, if I didn't speak up now, then I knew there was no hope. My dream of getting into space was rapidly disappearing so I actively took control of my education. I told them that I should be put in the upper stream and, with my home counties accent, they believed me. It was up to me to work as hard as I could to make sure I stayed there. I had sacrificed the happiness of a person dear to me to get to this point, and I needed to ensure that that sacrifice was not in vain.

I started doing better in science, and with a lot of effort and some relatively stable schooling, I began to excel. It was helped by wonderful teachers who would go the extra mile, and studying the sciences with my father.

We would often go to the Foyles bookshop on Charing Cross Road: a fantastic place spread over many floors in two separate buildings, a true treasure trove for any book lover. Father and I would look in awe at the books. It was hallowed ground, because he had started going there when he first arrived in the UK and it represented somewhere special. Father, Gracie and I also went to Swiss Cottage library, walking 45 minutes or so to borrow books on physics and other subjects, something we enjoyed doing together. It felt like a pilgrimage. As my aptitude in the sciences grew, so it did in other lessons because I was building more of

CHAPTER 3

an interest in school. *Maybe I'm not that stupid after all*, I thought.

I threw myself into my schoolwork because I had risked everything and sacrificed my relationship with my mum for this opportunity, so I couldn't contemplate failure. I think this spurred me on. It certainly meant that for the two years while studying for my GCSEs I didn't watch TV, and that was when *Dallas* was on. This was the one show everyone was talking about. I steered clear of the television and kept my focus on my reference books.

My daughter is neurodiverse, like me, and as a result she has hyperfixations. She asked me recently if I had any when I was her age. 'Education,' I said, which is not what a child wants to hear, but it was true. I wasn't going to shilly-shally around wasting time on pointless activities. I spent every spare moment at the library. One of my maths teachers, Miss Canning, even pulled me up on it. She said it wasn't fair on my sister Gracie, because I was so studious that it made it harder for her. The implication was that I was a hard act to follow, and it was somehow impacting Gracie's schooling. I think the teacher may have even told me to watch telly. I know other teachers thought I should loosen up, but nobody knew what my life had been like before. Had they known I had been in the bottom set and was doing everything I could to prevent myself from sliding back there again, they might have understood my motivation. I have been told off a few times in my life for working too hard, but that's why I love running my own company and working for myself – I can see the direct benefit from the work I put in.

Another reason my ability was taken on my word was because of my school history. Everyone assumed, after a run of boarding and convent schools, that I was bound to be bright. Added to my plummy accent, I just fitted their image of what a clever student looked like. Perception is an interesting thing.

At LSU, I had a few friends and was around long enough for these friendships to mature. My best friend was Helen Masterson. She was practical and often looked after her younger siblings. We formed a posse which included Susan Walsh (who had plans to become a major industrialist and offered to give me a job as an engineer later); Deirdre, a fey girl full of light and magic; Avril, a down-to-earth girl; and Margarita, a fellow physics fan. Margarita and I later ended up at Imperial together.

At the time, I was too focused on work to be interested in boys, although the gang and I were surrounded by chat about girls who had lost their virginity. We were at an all-girls school next to an all-boys school, and there were several crossover classes, so some of them fraternised. I thought it was a lot of hot air, but on GCSE results day, there was a girl from my class with a pram. I thought the baby was her young brother, but she said no, it was her child.

It was amazing that I did as well as I did because there were huge gaps in my early education, and even now

CHAPTER 3

I discover bits that are missing, like the Second World War for example. I never studied it. My father and I would always study maths together, and now I love helping Lori with her homework. I find it fascinating, and I'm always on the lookout for things I don't know so I can educate myself. I see maths problems as a challenge, something to get your teeth into and a wonderful feeling of achievement when you have conquered them, but Lori sees them more as a chore, so I try to change her response to them. We need to look at maths as a puzzle or game, like sudoku, and boost the element of fun.

I achieved good GCSEs, including English language and literature (let's not talk about French), which amazed me because they were tricky with undiagnosed dyslexia. I chose to do maths, chemistry, physics and biology A levels with art GCSE on the side. My father tried to talk me out of art because he said all my focus should be on the sciences, but I was resolute.

*

When I say I did nothing but work for my GCSEs, that isn't strictly true. I had one distraction. In the winter, walking home from school across Hampstead Heath when it was dark, I never looked where I was going because I was always looking up at the sky. What I really needed was a telescope.

DUNCE DUNCE DOUBLE D

My first telescope was a little plastic one from Argos. It didn't work very well, suffering from something called chromatic aberration. This means that as the light passes through the plastic lenses the resulting image gets split into a slightly overlapping red, blue and green view.

I was flicking through a magazine on evening classes and spotted a telescope-making class at a local school, Acland Burghley. I couldn't believe my eyes! There was no way I was going to miss this opportunity, so I went along to find out more.

I was 14 years old and walked into a classroom full of people in their 40s at least. They were all white and they were all male, but I didn't feel uncomfortable because we were connected by the same goal – the creation of a telescope through the laborious and difficult process of grinding and polishing your own mirror. Making one's own telescope is a process that can take years, and many people in the class were repeat offenders: serial telescope makers.

As well as the practical element of the class, we would go on visits to Hampstead's public telescope and view the cosmos in far greater detail. Even with the light pollution in London it was amazing what you could see, and if you got a really clear night it made for brilliant observation and a good ogle at Saturn and other planets. The telescope had to be controlled by our teacher because it was a large one with computer tracking and housed within a dome: not the sort of thing you can just have a go on without knowing

CHAPTER 3

what you're doing. But there are a number of astronomical societies who will invite people to come along at certain times of the year for a spot of stargazing, so it is more accessible than you may think.

When most people join a telescope-making class, they choose a Newtonian telescope because it's a simple one to start with.[4] These have a primary mirror and then a secondary flat mirror that you can buy. With this type of telescope, you are effectively looking through the wrong end of it because the mirror is at the bottom, and you look through where the light comes in. When the light hits the main mirror, it gets reflected up and then shunted outwards at the top of the telescope. I thought, *I don't want to do that.*

Instead, I wanted to make something smaller, more compact and trickier, called a Cassegrain telescope. You still have a primary mirror, but the secondary isn't a flat mirror, it is a very complicated shape called a hyperbolic. When the light comes in, it's reflected off the secondary mirror which

---

4. I didn't know it then, but Isaac Newton was to pop up in my future, or at least his telescope was. As an adult, I was at the Royal Society, the oldest scientific society in the world, for a photo shoot, and we were all trying to get into position, but there was an object in the way so without thinking I moved it. Someone said, 'Careful! That's Sir Isaac Newton's second telescope!' I wish they hadn't told me that because it was still in my hands, and I panicked. Thankfully I didn't drop it.

then sends light back through the end of the telescope, so you look through it the way you expect to.

All the big telescopes around the world are made to this design, and it means they can be much shorter. Newtonian telescopes can be a metre or so in length, depending on how much magnification you want, but I wanted something I could run around with, so the Cassegrain was ideal. This form takes the light and folds it, achieving the same magnification with something half the size.

It was foolish to pursue the harder process; starting with the simpler Newtonian would have made more sense. To make the primary mirror of the telescope, you start with two slabs of glass, put an abrasive powder in between them and rub them together. Because your movements are random, you end up with two spherical surfaces, one concave and one convex. Both are pockmarked, so you have to use finer and finer powders until you end up with a smooth surface. The problem is that a spherical mirror isn't very good at bringing light to a sharp focus. What you need is a mirror shaped into a parabola. When light from a distant object hits a reflective parabolic surface, the parabola brings it to a sharp focus. Perfect for observing.

So, once you have your spherical surface, the next stage is to modify its shape into a parabola. This is mainly achieved by working the centre of the mirror. This is complex because you are working the surface, sluffing off the smallest fractions of glass. The heat generated by the friction of rubbing

CHAPTER 3

the surface causes the glass to expand, so you cannot see if your shaping has worked until the glass cools down. If you start testing the mirror's shape too early, before it has cooled, then it gives the wrong reading. At this point, you need to test it by putting a beam of light through and obscuring half of the beam with a knife edge. This way you can see where the dips and bumps are on your mirror surface, and where you need to continue working.

In addition, the Cassegrain design requires you to punch a hole in the primary mirror. This is needed to send the light back up through the telescope to the eyepiece. The problem is that punching the hole causes stress in the glass, which introduces distortions that change the mirror's shape. When it is done, it is another hold-your-breath moment. At the end, you coat the surface of the glass in silver or aluminium, giving you the reflective mirror required.

What I have briefly explained here, in a few paragraphs, took me years to complete. I started at the class when I was studying for my GCSEs, continued through my A levels and I was still focused on it when I went to university, where it became part of my degree, so it probably took a good seven years or so to complete and refine. There were several points when I wished I had chosen the Newtonian telescope instead, but I persevered. As time wore on, I gained access to more sophisticated equipment so I could test it in better ways.

DUNCE DUNCE DOUBLE D

My telescope's debut was a moment of trepidation. The first thing I had to look at was the moon because it is closest to my heart. I couldn't quite believe that I was looking up at it through something I had made with my own hands, particularly after the amount of labour that had gone into it. The images were so much better than I had hoped for, but still I took the telescope apart and refined it a bit more. Talking of hyperfixations, this was definitely one of mine.

Chapter 4
The Chameleon

My father lost his eyesight when I was still at school. It had been a gradual decline, but we didn't realise how bad it was until one night we were in the car with him, and he almost hit someone as they crossed the road. We screamed out for him to stop, and he braked just in time. We sat in silence for a moment.

Some time before that near miss, Father had been diagnosed with glaucoma and was on the list for an operation, but there seemed to be an assumption that because he travelled a lot he was a Nigerian playing the system, so it was cancelled. He was taking medication, but it didn't stop the decline.

As his sight began to fail, I took on more of the household duties alongside working for my GCSEs and then A levels. Every night we would fervently pray at the altar in Father's bedroom, and I would ask God to give him back his sight. I was constantly thinking of sacrifices I could make, things I could offer up to the Almighty that might convince him to make my father better. If I worked really hard,

CHAPTER 4

perhaps? If I was a very good person, maybe? If I didn't tread on the cracks in the pavement as I walked to and from school?

Father remained convinced that there were evil forces at play that were disrupting his sight, and as his vision decreased so his paranoia soared. He devised symbolic practices to protect us – like writing psalms on eggs and then eating them – and this engendered quite a bit of fear. He continually lectured us on the hostility of the outside world towards Black people, something I was already experiencing. One school friend, Sangeeta, told me if I was a boy and she took me home to meet her Asian parents, they would be horrified because of the colour of my skin. I felt this with boys too. Nobody wanted to go out with me, the Black kid with probably some rather funny mannerisms. At the time, the environment I was in was fairly white and it felt alienating to me, so I buried myself in my education and the local library.

I loved reading from an early age, even though dyslexia threw banana skins in my path. TV helped the cause. I watched a translated version of *Heidi* on the BBC. It was badly dubbed with the mouths not really corresponding with the voiceover, but that just added to the charm of it. I then discovered the book at school and, on reading it, I realised the joy of visualising something with my own imagination. Heidi could look any way I wanted her to look. The hayloft where she slept was how I envisioned it from the words on the page, not from what I had seen on the dramatisation. So, although I knew the story and loved the TV adaptation, reading the book made me go 'wow!' Suddenly, books became more interesting. The same happened with *The Secret Garden.* I watched it on telly and then picked up the book. It was slow progress, but I assumed this was because I wasn't working hard enough, and if I did then I would be like everybody else.

I gave a talk once, entitled 'How Science Fiction Saved My Life'. Most of the books I was told to read at school held no interest for me, but when I discovered science fiction on the page as well as on the screen, there was no stopping me.

John Wyndham's *Chocky* was one of the first I read, and I was transported. I devoured it, sitting in my mum's car while she was organising something in the village hall near Hastings. It is about a 12-year-old boy who has an alien who speaks to him through telepathy and can see through his eyes. It's quite anthropological because the alien is commenting on the human world, which creates conflict between it and Chocky. It appealed to my obsessive nature. Once I start a book and I know it is good, I can't put it down.

Towards the end of school, I began to read science fiction written by women, which was a welcome relief after some of the more misogynistic approaches. I also discovered Julian May (a woman), who wrote a fantasy sci-fi series about people with psychic abilities who connect to the galactic community. This storyline really appealed to me with the idea that we could communicate with aliens and that they would introduce us to their technology.

CHAPTER 4

As well as taking books out of the library, I borrowed records too. I got Bach's Brandenburg Concertos, and I put it on the record deck at home, listening to it with tears welling up.

The first gig I ever went to was accompanying Gracie to see Rick Astley. I wasn't much of a fan, but Gracie thought he was fab. I enjoyed the concert much more than I thought I would because he was brilliant live. While I was at university, my old school friend Helen invited me to come to Reading Festival with her and we wandered around listening to different bands. Most recently I was invited by Professor Alice Roberts to join her at Glastonbury, to take part in one of the literary sessions. I took Lori with me in our camper van. The number of people there was terrifying, but we had so much fun. Thank you, Alice.

Music fascinates me: the playing of a chord and how it can make you feel, whether because of the sound itself or the memory it stirs in you. Do I react to this piece of music because that is how it is designed to make me feel, or am I responding to something in myself that recognises it, that links it with a nostalgic memory? I don't yet have an answer to this, but many people are studying the psychology of music.

*

THE CHAMELEON

My father's eyesight continued to degrade. By now, he could just about make out dark and light shadows, but all other detail was lost to him. One day, I looked into his eyes and realised his iris was ruptured. The glaucoma had literally torn his eyes apart. It was so severe – far worse than I had ever thought – and in that moment I knew it would truly take a miracle for him to get his sight back.

Even now, the emotions that surge inside me catch me out, and I have to pause to regain my composure. I am so sad for what my father lost.

He was amazing at getting about, and was initially determined that his gradual loss of sight would not hold him back. He was given a white stick and would make forays into the world with it, catching trains and finding his way around independently, occasionally relying on the help of a stranger. With his diminishing sight there seemed to be less fear of racism. His independence used to terrify and amaze me in equal measure. When I was at university we would occasionally go out for meals. One unfortunate moment is burned into my memory. Father still had some sight, although not much. It was my birthday, and we went to Debenhams; I was feeling sophisticated after a new haircut, so was rather distracted and didn't keep an eye on him. He tripped on some steps. He was mortified. We both blamed me for my moment of distraction, and I still carry our shared disappointment.

CHAPTER 4

As the disease took more of a hold, it also stole more and more away from him. My father was hugely independent, a larger-than-life character and a happy, brave global traveller. As a child I remember a black-and-white picture of him sitting on a camel, wearing a fez. But in the end he was reduced to shuffling around our three-bedroom council flat, being looked after by his teenage daughter and living off disability benefits. He gradually withdrew from society.

I took on more responsibility at home. I would collect Father's disability benefits from the post office. I would use this to do the household shopping on a tight budget, cleaning and cooking for the three of us. I didn't think of myself as a carer at the time. It wasn't a label that was used back in the 1980s, but recently I can see that I was – without the support or the infrastructure that we needed.

In some ways, I was a child taking on the role of an adult with the possible jeopardy this entailed. We had an old cooker, and the gas regulator didn't work very well. Once, I lit it with matches without knowing that the gas had built up and *whoosh!* There was a pungent smell of burning hair and I was missing part of my eyebrow. I remember thinking we would need to get someone in to sort out the stove. A while later, I was lying on my bed when I heard exactly the same noise. Realising my father was lighting the cooker, I sprang up and ran into the kitchen, convinced I would find him lying on the floor. Thankfully he was fine, and, after speaking with the council, I got rid of the damn cooker and bought a second-hand one.

One of the long-term effects of his blindness was that he didn't move around very much. Initially we would get out and about but, over the years, and with further loss of his sight, his mobility was much reduced. As a result, he got horrendously painful leg ulcers, creating this vicious circle where greater mobility would have helped but the excruciating pain associated with walking meant that it was not an option.

Looking back, I wish I had taken him out much more, leaning in to those early days when he still loved his independence even though he was blind. I loved the tales he would tell about his interactions with kind strangers who were more than happy to help such a jovial, smartly dressed blind man.

By the time Gracie and I left Hastings and joined Father in London, Sue and Hal had left home, graduated from university and were settling into their careers, so they would only visit occasionally. There was still friction because Father felt they had somehow let him down, and as a result he had partially cut them off. He had a habit of estranging people. Gracie and I would sometimes go and see Sue and Hal, but most of the time it was just the three of us.

During this dark time, he found one ray of light. He was determined to study law and so we would get books from the library; through the Royal National Institute of Blind

CHAPTER 4

People (RNIB), we found people to read them to him. They would record the books onto a cassette, and he would spend days listening to them, so even though so much of his life was diminishing, he was still hungry for knowledge. He was a fighter.

Sometimes people would come to the flat to read to Father, and he would record it. It was also company for him. One young woman came regularly, and they got on very well until tragedy struck, and she was killed in the *Marchioness* disaster, when a Thames riverboat sank. Her parents visited my father at our home and told him how much their daughter had enjoyed reading to him. I remember hugging them, but being unable to convey in words my deep sorrow for the loss of their kind daughter.

I never felt that looking after my father was an imposition – our bond was too close for that – and I didn't use education as an escape: as I said, it was more of an obsession. At sixteen, I felt I was a grown-up. I was in charge of the house and had independence. One of the things I wanted most in the world was my father's company, and Gracie's for that matter; now we were all together, and doing a few chores was not going to undermine that. I didn't have time for the teenage angst that came with friends, boyfriends and social events, because I never went out. I have done this a few times in my life, focusing inward rather than looking outward for company. When Lori was born, she too became my focus and I had little time for socialising as she grew. Only now, as she enters her teenage years and grows in independence, do I realise what I might have missed out on as a child.

*

La Sainte Union Convent School turned out to be a great place for my fierce academic determination, and my time there was the longest I spent at any school. Years later I think they named one of the labs after me, which was lovely and reflected the affection and gratitude I felt towards the school.

It did seem mad to take four A levels, but each subject gave me the information I craved. Physics gave me an understanding of the universe on all scales, chemistry many of the processes that govern it, biology the understanding of life – human, animal and plant – and maths was the language that enabled us to explain it all. How could I drop one when they all enabled us to understand life, the universe and everything, to borrow a phrase from Douglas Adams. This package also turned out to be great for my later calling as a science communicator.

A sixth former used to come over from the boys' school across the road to study A level biology with us under our teacher Mr Halstead. I can't remember the boy's name, but he was my prime competition. I think this stems from me being determined to be as good as any boy, and while he was my nemesis, the healthy competitive spirit drove us both to work harder.

CHAPTER 4

On the flip side, I had to go to the boys' school with my classmate Margarita for physics A level, which was a different examining board from the one I had for GCSE, although I enjoyed the rigour of the subject. Our first homework assignment was to guesstimate the area of hair on someone's head, and I came up with two methods. The first was to lay paper over the head and get an area that way, and the second was to estimate the surface area of a sphere the size of a human head and then estimate the percentage that would be covered with hair. Margarita and I shared our dismay at the very different approach to physics questions with the new exam board and encouraged each other. She chose the paper method and I the spherical head estimation, which seemed more scientific to me. However, she was praised by the teacher for an innovative approach.

We missed our GCSE physics teacher Mr Vesty, who was brilliant and made science a real adventure. Another teacher at the boys' school was Mr Detheridge. He had worked with Michael Nelkon (of Nelkon and Parker, the authors of key physics textbooks that my father had had at school) and he gave me extra lessons.

My maths teachers were unsure if I should continue with maths A level, because I would get my numbers the wrong way round and sometimes put them in a muddled order. The decider to my fate was an exam we all took called an AS level: something in between GCSE and A level. I did far better than expected, and was allowed to continue.

THE CHAMELEON

I got a B, two Cs and a D. I got a respectable C in GCSE Art and I am still glad I took it. I am usually quite busy but one of my favourite ways of relaxing is doing art projects with Lori. The D was in physics and surprising considering the support I received from Mr Detheridge, but I think taking the four subjects was a bit too much of a stretch. I sometimes refer to this in my public speaking or when I meet a student who is devastated by a low grade. I always tell them that success isn't about not failing. It's about how you pick yourself up when you do fail – and my life has been littered with potholes like that. I still got to where I wanted to go, through university, in work and in life. These exams do not define us, and a slip-up doesn't mean the end of one's dreams; it often just means that we need to find a different route to get there or, in some cases, find a different destination.

I applied for a place at Oxford as my father was keen for me to do, but I didn't pass the written test. Knowing what I do now about my dyslexia, it makes total sense, though of course there may have been other factors too – with my education, was I really Oxbridge material?

While physics was my worst subject on paper, I still chose to study it over biology and chemistry at university. As I've said, to me, physics is the study of everything in the universe, from the outer reaches of the cosmos to the tiniest subatomic particles; every chemical reaction is based on a physical process. With physics, you have the fundamental knowledge for all sciences; it feeds into everything else.

CHAPTER 4

Father wasn't sure about my decision to do a physics degree because he still wanted me to study medicine. He couldn't understand what I would do with the degree once I had it, but I had made up my mind and I went for an interview at Imperial College London, my first choice. When I received the offer from them, I was ecstatic.

Strangely, Margarita also got a place on the same course, which really threw me. I had a habit of reinventing myself with every educational move I made, taking on a different form, the chameleon at work. I had been at LSU for four years with slow incremental change. Now, when it was time to metamorphose for the first time in a long time, I was starting somewhere new with someone who already knew me. I couldn't change who I was, shed my old skin and move into a different era, because Margarita was my witness. She knew the old Maggie, so how could I be a new one? I almost resented her for unwittingly limiting my growth.

From gritty inner-city primaries and unauthorised absences to private boarding and wholesome coastal schools, I experienced a multiplicity of different educations all woven together. I lived almost every version of childhood. Even then, I saw this as a superpower, a strength of character to be robust enough to adapt to each move and new school. This may sound surprising for someone so sensitive, but I think it was the making of me. I was forced into new situations regularly and had to sink or swim. It toughened up the small, vulnerable, empathetic child into something more resilient.

THE CHAMELEON

I still think of myself as a chameleon. I can fit easily into new environments and adapt to what is around me. I think this is an amazing coping mechanism so, while the changes are less dramatic for me now, I can flex that muscle when I need to. In my late 50s, I am again going through a time of change. I have been a wife for over 20 years and a mother for 15, but now Lori is a teenager, so does not need me in the same ways she did before.

I have evolved with her and hopefully adapted to her changing needs – and as she grows in independence, I am not frightened. I am excited! I think I am ready.

*

I didn't have much of a clue about clothes at school, nor did I have the money to do anything about it if I had wanted to. I must have been a scruffy mutt, because one of my teachers pulled a friend to one side and asked her to take me shopping to buy an appropriate outfit for university interviews. So, once I had accepted a place at Imperial, I knew I had to plan my wardrobe.

This was the first time I had to actively think about what to wear. To date it had mainly been school uniforms and, at the weekend, hand-me-downs from Sue and Hal. I needed to take into account budget, ease of dressing and my severe lack of interest in fashion. I went to a place in Camden called

CHAPTER 4

Lawrence Corner, a treasure trove of surplus military clothing and other delights. I had brought a full-length military coat from there in the past. It looked quite Russian to me, and I would wander around in it, imagining myself as Yuri Gagarin, Valentina Tereshkova or some other cosmonaut.

Physics at Imperial was going to be hard, so why get distracted by clothing? So, for university, I bought two military jumpers in navy, which came complete with elbow patches and shoulder straps for stowing a beret. I admit this sounds like a strange sartorial choice, let alone buying two of them, but I figured I needed a practical uniform, and the jumpers fitted with my image of that – one on and one in the wash – and they were inexpensive and looked as if they would last well. I thought I would wear the same outfit every day because my focus was on science. It's funny to think of this because I am now *obsessed* with clothes, but until I finished my first degree I thought this was just frippery, a silly distraction from the serious business of education.

Perhaps some of this dressing down came from the idea that 'dolling oneself up' was a waste of time, and also something that my father associated with my mum and so should be avoided. I was brought up with the notion that paying too much attention to one's appearance was at best a distraction and at worst a way of taking advantage of people – using one's feminine wiles to get ahead. The implication was that my mother, who is very beautiful, did this, and that I should be very wary of doing the same, even with my limited looks.

In some ways, life didn't change much when I went to university, other than having further to travel every day because I had to go to South Kensington. It felt like an extension of school. I still lived at home; I was wearing my own version of a uniform; and I couldn't regenerate into the new Maggie in front of my old schoolmate. Almost everything was familiar, except the workload. If I thought A levels had pushed me, then this was a whole new level of academia.

My first year was pretty brutal because we had to get up to speed with maths, which the lecturers explained was a big leap from the subject we had done at school. They were pulling us up by our bootstraps and there was no time to waste. It was like learning a different language made up of numbers and squiggles.

In those early months I did very little outside of my studies, although for some inexplicable reason in freshers' week I did a test run with the gun club. I wanted to know what it felt like to fire a gun, but the second I had a go it terrified me. I released two shots and realised that if I made a mistake, I could kill someone by accident just with the movement of a finger. Where was the fun in that? No, thank you. I left and didn't join any more clubs.

My highlights of that first year included doing more hands-on lab work, which I loved, and being taught by some fantastic lecturers, like Professor David Southwood, my tutor, who was so warm and friendly: the ideal person to start you off in uni life. It turned out he was a space scientist,

CHAPTER 4

and he was the head of the group at Imperial. He would later be a director at the European Space Agency (ESA). I had struck gold. At that stage, I didn't tell anyone about my dream to be an astronaut and go into space because it still seemed like a crazy idea, so I kept it my most guarded secret. I didn't tell David either, but it felt like fate had brought us together, and his encouraging presence was a big part of building my confidence. Another member of my tutor group was one of the few other women on the course. Her name was Nicola Fox and she has become head of science at NASA, so David's influence was strong.

*

My first-year project at Imperial was focused on the telescope I had started making at evening classes. I wanted to make a little wooden box to house the mirrors and work out how to support them, practices which would stand me in good stead later as a space scientist and astronomer. There were challenges to overcome, which I loved. I was lucky that I could bring a project in that I had already been focused on for several years. In the summer holiday, I was allowed to come into the workshops and continue working on it. I learned many skills and people were so kind and accommodating in sharing their knowledge and giving me space to work.

THE CHAMELEON

While I was at Imperial, I also made a computerised tracking system for my telescope. This is a system that allows you to select an object to observe, for instance a star or planet, then if it is visible in the night sky, the system will automatically move the telescope to view the object. Also, if you wanted to observe an object for a prolonged period of time, because of the Earth's rotation the object will drift in the field of view of the telescope until it is lost completely. The tracking system was designed to keep the object in sight as the Earth moves.

Nice idea, but to do this you need a starting point, a known point in the sky from which you can track. At the time when I was designing the system, we didn't have GPS on our phones. So, I came up with the idea of having a plumb line attached to the telescope. A plumb line gives you a vertical direction. If you set up your telescope in alignment with the plumb line, then it is looking directly upwards. With a simple calculation using the time of day and your position on the planet, the tracking system can calculate the night sky directly above your location. This gives a fixed reference point, which is my starting point to track from. I installed motors and gearboxes into the telescope housing so the tracking system could navigate the telescope across the night sky.

When I finally finished my telescope, I wondered if I would be sad or if there would be a clanging feeling of anticlimax. In fact, I realised that in taking on something so

CHAPTER 4

epic, it was more about the journey than the end result. I also knew I could continue to tinker with it if I wanted to, so it was never quite the end. I have bought a professional telescope, but the truth is I don't use either of them very often because I don't have much time. Nor have I made another telescope, although I like the idea of making a Newtonian one with Lori. It would be a nice project to do with her.

When people interested in astronomy ask me what they should buy, I always say if you are really dedicated and also want to do astrophotography, then go for a telescope. If you are a more casual user, I recommend astronomical binoculars, because there is virtually no set-up with these. You simply shove them around your neck, step outside and you have immediate access to the night sky. Although you won't get as much detail as with a telescope, you can still see so much through them, and with none of the setting up.

Recently, I returned to the telescope-making class as part of an episode of the BBC's *The Sky at Night*, which I co-present. Can you believe the same teacher was still there?! I had also included it in a book I wrote about the moon, and one of the students came up to me that night and said they were there because they had read about my experience and wanted to make their own telescope. That was quite something to hear.

*

THE CHAMELEON

While incorporating my telescope in my first-year project was a symbiotic experience for me, it also meant I got to use the BBC BASIC computers and high-density floppy disks, as opposed to the Amstrad we had at home. It all seems like such archaic software now, but at the time this was cutting edge, until an even newer computer arrived, the Archimedes, with its nice flash screen and its speedy processing skills. It was powerful. I always imagined it with a glowing aurora around it. There were many BBC BASIC computers, but only a couple of Archimedes machines, so they were coveted items in the lab. The stargazing software I was developing would take a few minutes to load on the BBC machines but the Archimedes would display everything in less than a second.

Studying for a physics degree was a lot more intense than many other subjects; with lectures, laboratory practicals and homework, there was very little downtime. The little down time I did have was partly spent on the Pimlico Connection programme, which was run by Imperial. As undergraduate students, we would go out to local schools and encourage the kids to consider careers in STEM. This would be my first foray into science communication.

At Imperial, most people were in the same boat because they were studying either sciences or engineering. My timetable was similar to the one I had at school, with no gaps to waft around or time to flop on the grass by the tower that dominated the university building. I barely came up for

CHAPTER 4

air for two years. What with my academic work and still living at home, I knew I wasn't getting the fully rounded student experience. I wanted to get more life experience, but didn't quite know how.

I needed an adventure.

I hatched a plan with my then lab partner, Gideon. We were students and we wanted to travel, so we thought why not get funding to go somewhere amazing and do some experiments. After much debate, we chose the Amazon. I mean, if you can go anywhere in the world, why not the Amazon? We were slightly on the back foot with this choice, though. The Amazon lends itself well to biological studies but slightly less so to physics-related investigations. Looking back, there were some amazing astronomical sites we could have gone for instead, but we wanted to travel, to see the world, to go to places off the beaten track. Be brave! Be free! Be intrepid! Why not apply for funding to go on an expedition to the Amazon rainforest? No matter that we were physicists who were being distracted by biology research – surely somebody would fund us?

We put a poster up in the student union looking for volunteers to join us. Two people applied: Sue, an electrical engineer, and Baz, a physicist like me. But we needed biological scientists, not more physics and engineering students, so we declined their approach. Instead we signed up some biologists – who soon realised we were non-essential for the project and ditched us. However, we became close friends with Sue and Baz, who later got married, and still laugh about this with them some 30 years later. Sue's lab partner at university was my future husband, Martin, and I am so glad that we have all kept in touch throughout our lives.

Our plans grew bigger and more convoluted and then, just like a balloon filled with too much air, they burst before we had time to find our passports. Still keen for an adventure, though, Gideon and I decided to go interrailing instead. I had been abroad once before, on a ski trip to Italy with school when I lived with my mum. It had been a lovely experience, but as it was a school trip with my teachers, I did not feel very independent. I was desperate to leave the UK and see the world just as Father had done.

My father both inspired and discouraged this decision. Before he lost his eyesight, he had travelled extensively and would tell me tales from far-flung lands, filling my head with exotic sights and the desire to see these places for myself. When I told him I was going interrailing, though, he was less than keen. Hadn't he told me of the dangers? Didn't I realise how vulnerable I was? A young Black woman in a hostile world? He wanted me to be safe but, in that desire, he weighed me down with apprehension. I knew if I let him stop me I would never spread my wings. Hal came to see me off on my journey, assuring me of my decision. She had recently returned from a trip from London to South Africa by truck, and was part of the inspiration for my small adventure.

CHAPTER 4

Gideon and I left the UK with a planned route to France and then on to Italy, Greece and Morocco. My rucksack was packed so full, every time I put it on I almost fell backwards. I have always been a terrible over-packer. On and off trains, interacting with strangers, I was absorbed in every new experience. At one point, we were on a boat for an overnight crossing from Italy to Greece, with tickets making the ferry free but only as deck class. So rather than pay extra to get a cabin, we slept on the deck, warmed by the Mediterranean night air and watched over by a multitude of stars.

I kept my rucksack and my money belt close to me always. I was incredibly wary and could hear my father's voice in my head. In many ways this was useful, as it taught me to be cautious.

Early on in our journey, I think travelling through Italy, we were sitting in a carriage and a guy came in, asking if he could join us. After a little while he told us he needed to nip out briefly and that he would just leave his bag, but there was something about him that made me feel uncomfortable. I said no, although it seemed unbelievably rude to do so, but he didn't take offence and moved on with his bag. Sure enough, a few minutes later security came through looking through everyone's bags. This may have been completely innocent, but I did not want to try and explain, in a different language, that the bag in our cabin actually belonged to someone else. From a very sheltered childhood, I was learning when it was safe to trust people, and when it absolutely wasn't.

THE CHAMELEON

In Greece one evening, I had a bit of an argument with Gideon and stomped off out of the restaurant on my own. It was quite late and a rash thing to do. As I left a man followed me outside and fell in step with me. We chatted for a few moments about nothing in particular and then he reached out, putting his hand over my mouth. I was totally alone and couldn't see anyone nearby. I felt a jab of fear, but instinctively I smacked his hand away and shouted, 'Don't do that!' He was so stunned he stopped, and we kept on walking together till we got back to the restaurant. Looking back, I should have been terrified, and I think I was, but in that moment, I was determined not to get clonked over the head or abducted. I could imagine my father shaking his head if I survived to tell the tale. I would not be a victim. My late-night forays in London certainly helped. If you look timid and scared, you can be seen as easy prey, whereas if you walk tough, even if it's just an act, I believe you are less likely to be bothered.

I was adamant that I wanted to go to Morocco because I had never been to Africa, and it was the closest I could get to the motherland on an interrail pass. Gideon was definitely up for it too. Why travel all this way and not clock up another continent?

It was twilight when we got off the train in an out-of-the-way place. We had been carrying some camping gear and this looked like a wonderful opportunity to use it, in the forests of Morocco. But as we began walking, slowed

CHAPTER 4

down by our backpacks, a group of men started following us. We realised that setting up camp under these circumstances would be at best foolhardy and worst deadly, so we doubled backed to the train station. The station master, a wonderful man, confirmed that we might not have seen morning if we had continued with our camping plans. As there was no hotel in the area he agreed to let us sleep in the station overnight. He was incredibly hospitable and made us mint tea just the way I liked it, sugary and vibrantly fresh. He was so kind, and a reassuring juxtaposition to what had been a threatening situation.

At the end of our four-week trip, we decided to head back into France and spend a week or so there before we had to return home. We had no money, but we had a vague plan to get a job working in a McDonald's (I'm not sure why, nor do I think they would have wanted me with my pidgin French – it was just a way to try to extend the journey).

We were on the train from Morocco, speeding through southern Spain, when I did something I still can't quite believe. I tucked my passport in its holder between the headrests of our seats, *to keep it safe*. I know. What can I say? I thought that was the best place for it. A while later, while Gideon was asleep, I wandered down the train to find the lavatories.

The first set of lavs I got to were disgusting. There was no way I was going to use them, so I kept going, but the door to the next carriage was locked. *Not a problem*, I thought. I would just wait for the train to pull into the station where

THE CHAMELEON

I would hop off and back on the next carriage to use a nicer loo. This seemed like a good plan at the time because at this end of the train I was getting a little too much attention from a gang of boys who were looking at my money belt. So as the train slowed I jumped off. But I misjudged the speed of the train; it looked relatively slow from the carriage but, on hitting the ground I realised it was still going fast and I rolled to the ground, in a cloud of dust. I picked myself up, glad to note I was not seriously damaged, and brushed myself off with relief, only to realise the train wasn't stopping. *GASP!* It had slowed down to go through the station safely, but it wasn't the next stop.

The train picked up speed and I began to run after it until I gave up and watched it disappear into the distance, with Gideon sleeping soundly and none the wiser. What on earth would he think had happened to me? I was still wearing my money belt, so I had some cash on me, but then I remembered my passport in its holder was tucked 'safely' between the seats and in that holder was also my train ticket.

I cried out in pain when I realised what had happened. I wanted to sit down and weep and I think a few tears did flow. I pulled myself together, telling myself that sort of behaviour wasn't going to solve anything. Again, I could visualise my father sadly shaking his head and saying, 'Oh Margaret, I told you so!' However, I was not going to give up without a fight. I tried to explain my predicament to the train staff, but what with the language barrier and the fact

CHAPTER 4

they had just witnessed me throw myself from a moving train, they weren't terribly interested. The only thing for it was to get on the next train heading to the same destination and hope that was soon. (Can I also remind you this was in the late 1980s, before everyone had mobile phones, lest anyone wonder why I didn't just call or text Gideon.)

I had no passport, no ticket and the only clothes I had were the T-shirt and leggings I was wearing. And I was still three days away from Paris.

My new journey began. I was on and off trains, attempting to explain my ticket issues with every inspector that came along. In the worst of times, the kindness of strangers really does shine through, and I experienced this. After witnessing my regular sobbing episodes, a bloke I was sitting next to tried to piece my story together, even though he couldn't speak English, and I couldn't speak French. A German passenger who spoke both languages translated for us, and they both spoke on my behalf when I needed it.

I was already scrawny after a month of walking around carrying my belongings on my back and living off biscuits because of our limited budget, so people offered me food. My biggest fear was that the authorities would assume I was an illegal immigrant, trying to get into the country from Africa, and I had no documentation to prove otherwise.

Amazingly, I managed to get all the way to Paris and there went straight to the police point in the train station to explain the situation. I half expected to be deported, but as

THE CHAMELEON

luck would have it, Gideon had found my passport and left it with them along with an address for where he was staying. I think all sorts of theories had run through his head over the few days we were apart, starting with murder and abduction, but he was a sensible chap. When we eventually met up, we hugged, laughed and I kept apologising. I kept the hugging quite brief because I hadn't changed my clothes for three days and was probably a bit whiffy.

I learned several valuable and obvious lessons from this experience (besides 'don't hop off trains until they come to a complete stop'). One of the biggest was realising I didn't fall to pieces in the face of a crisis. I could rely on myself. Yes, I also relied on the kindness of strangers, but being able to do so was in itself another win. It gave me an independence I did not know I was capable of, after a lifetime of being convinced I was vulnerable and unable to look after myself. Even though I had been running a household from the age of fourteen, I thought I couldn't do anything without other people to support me.

*

The interrailing trip irrevocably shifted something in me. When I returned home, I felt fidgety, like a wild pony trapped in a small stable. I was 20, I was going into my third year of uni and I wanted to truly experience university

CHAPTER 4

life while I still could. To work out who I was and what I was capable of. The interrailing was an opportunity to see myself in action and rely on my wits and, I must admit, I wanted more.

Gideon was living in a tiny bedsit in a large house in Philbeach Gardens, near Earl's Court, and he mentioned that another room had become available. When I say bedsit, it was just big enough for a bed and a sink and not much else, and there was a shared bathroom on each floor. The landlord was called Mr Christian, and he owned lots of bedsits in the area, so we nicknamed the houses he managed the Christian Empire. It meant I would be much closer to uni than being at home. I had to work out if my student grant would cover it, but the main issue was breaking it to my father. How could I tell him I was moving out? Surely he would understand – after all, I was still in London. I would be at home less but I would still come round and help out.

He did not. He was absolutely furious. I got the full force of his Nigerian tendency to dramatise the situation, as he told me I was no longer his daughter, he was disowning me and banishing me from the house, never to set foot inside again. I watched our relationship implode in front of my eyes. I couldn't believe he was doing this; we had been through so much together, he was my linchpin and now he was throwing me out and cutting me off.

At the time I was devastated. Yet I realised he had effectively done the same to my two older sisters when they became independent. I guess I shouldn't have expected it to be any different for me, but it hurt with an indescribable pain, the remnants of which I still feel today. To my horror, I had become one of the 'witches' who had 'defied' him.

I could have reversed my decision, but I knew if I didn't leave then, I never would. I saw that very clearly. That foresight gave me a sort of strength. There was guilt, as I was putting my needs first, but with a less brutal break I think that the change would have been minimised. Instead I was out in the world for the first time on my own, with no real idea of how to cope. I knew I could look after myself on a budget, I had been doing that for our small family for years, but emotionally it was a very different thing.

One of the upshots of my unceremonious dismissal was that I was not just leaving my father, but also Gracie who would now be in charge of the household; it would be just her and Father. She and I didn't talk about it until many years later. She told me how betrayed she had felt, that I had seemed to leave without any care for her. I did try to explain that this was not what I had planned. If the break had been closer to what I had envisaged, it would have been less severe and she would have felt less abandoned.

But I had made my choice: I would leave the warm, cosy family nest and step into the cold, unforgiving world. Father's warnings about the worst things happening reverberated in my head as I navigated those early weeks, living alone in a tiny room. Without contact with my family

CHAPTER 4

I felt incredibly lonely, vulnerable and disconnected from my past, so I once again buried myself in student life.

I didn't sail through university, and neither would I say I survived on gritty determination; it was a fight, and I didn't always get the grades. I slogged through three years and came out with a 2:2, which I wish could have been better. With the retrospectroscope, going independent in my final year may not have helped my grades, but it felt like a bid for freedom and a 'now or never' sort of scenario.

I went back to Imperial a couple of years ago, to film a segment for *The Sky at Night*, and met up with Nicola Fox. It was brilliant to see her, and she was just as fun and witty as I remembered her. It was strange to return but I enjoyed walking around college with her, reminiscing, visiting the laboratories and the lecture theatre, which hasn't changed in 30 years. We could even remember where we sat. I would be in the second or third row, two seats in from the aisle so I could pay attention. They had unearthed some old photos taken during our time there. I trawled through to see if I could spot myself, but the only person I recognised was Gideon. They did have one picture of me from that time: a head-and-shoulders shot of me wearing my bloody military jumper!

Chapter 5
Doctor, Doctor

For much of my life, I had this idea of being an academic. I would visualise myself with grey hair and a pipe, thinking deep thoughts, writing a few complex but meaningful equations on a blackboard, and suddenly the universe would make sense. The reality was quite different. My degree had highlighted some of my strengths and weaknesses. It was clear that I was better at the hands-on practical exercises than I was with my head in my textbooks. Maths was OK, but long reams of prose, the sort of thing you get in academic papers, addled my brain, which was a problem because academics live on the written word. They are judged on the papers they write about their research, and their funding relies on them. While I wasn't ready to give up on academia just yet, I needed to find the right fit for me.

I may have got a degree, but did I think this was enough to be taken seriously? No. What I really wanted was a PhD, to be a doctor, and my heart was set on achieving it. While getting a 2:2 for my degree was fine, it wouldn't grant me a government-sponsored research position; I needed a 2:1 for

CHAPTER 5

that. The usual route was to do an MSc (a master's in science) next, but that would take another year or so and I wanted to find a quicker way. ('Impatience' should be one of my middle names.) Furthermore, I had been running up the educational hill for some time now. Four years at school since leaving Hastings, another three years at university to date. Most PhDs took around four years, so with an MSc that would be five. Would I have the momentum to keep going, or would I just stop after the MSc?

I had seen an advert on the physics department noticeboard at Imperial. The tribology section of Imperial's mechanical engineering department were looking for a physicist to join them to work on a privately funded PhD project. They were looking for someone who specialised in optics to work on a new machine that measured very thin films, and my work on the telescope had given me a good grounding in this area.

I made contact and the supervisor for the project, Professor Hugh Spikes, called me in for a meeting. I went from the familiarity of the physics department to the unknown world of mechanical engineering. In my three years at Imperial to date, I had never set foot in mech eng, as the department was called.

After I was accepted for the PhD in mech eng, the alienation continued. Even the measurements they used were different; in physics we use modern système international (SI) units: the metre for length, the kilogram for weight and Newtons for measuring force. As I was introduced to the kit I would be working on, people were talking in different units that I had not encountered before. Things like pounds per square inch. It felt archaic, like I had stepped back into the dark ages. What was next, furlongs per fortnight?! This was mind-boggling to me. It seemed odd, but I suppose it was just another kind of adaptation, and I was not going to let a few weird units stop me. It was still a really good fit for someone with my skill set.

I prepared myself for a change of direction. I was more excited by the opportunity it would present and intrigued at the use of optics for this challenge. The goal of a PhD as I saw it was to conduct some original research that would add to the wealth of human knowledge, which – put like that – is quite a lofty ideal.

The transfer from school pupil to university undergraduate is quite a big step, but some things remain familiar. One is being part of a large cohort of students with a timetable of lectures to attend, practicals to conduct and tutors to interact with. Now, for the PhD, I was conducting independent research, so the structure was very different. There were supervisors to guide me, including the wonderful Professor Philippa Cann, but much of what I did was off my own back. I remember one of my fellow PhD students, Rob Dwyer-Joyce, now a professor at the University of Sheffield, had a supervisor who was delightfully eccentric, as only academics can be. One day Rob went to him to ask him a question about his research. His supervisor, who

CHAPTER 5

had the quintessential pipe, took a thoughtful puff and then looked off into the distance. His response was simple. 'The answer,' he said, 'is in a book.' He paused for what I can only assume was dramatic effect and then added mysteriously, 'Somewhere.' We fellow students fell about laughing when Rob recounted this to us and 'in a book … somewhere' became a sort of meme in the lab. We would quote it to each other whenever it seemed appropriate.

After the shock of changing department wore off, I began to enjoy the anonymity of being somewhere new. Even though it was the same university, the departments were quite separate, so it was unlikely that I would have any major dealings with my cohort from physics. Change seemed familiar to me; it was another opportunity to take on a new persona. I'm not sure why I felt the need to leave almost everything of the old me behind before I metamorphosed, but I did. I wanted a clean slate. I moved away from military jumpers. It turned out interrailing for six weeks was the ultimate diet, and I had lost quite a bit of weight on my adventures. I didn't have much money to transform my wardrobe, but I discovered the young women's clothes shop, Miss Selfridge, and my interest in fashion was ignited.

There was another woman doing her PhD in the mechanical engineering lab (known as the 'mech eng lab' to its friends), but she was a few years ahead of me. By a strange coincidence, she was also called Margaret and she was also Black. What are the chances of that?! It started to cause some confusion, even though we were working on different projects. So, I adopted my family childhood nickname, Maggie, to make it easier for everyone. (Margaret had always had negative connotations to me anyway. If I was naughty as a child, my full name would be used in a stern voice, so I always preferred Maggie.)

The majority of my fellow PhD students had degrees in mechanical engineering, and we had a sprinkling of chemists among us, but I was the only physicist, so I regularly felt on the back foot regarding the ways of mechanical engineering. I only realised how aware my cohort was of my limitations when we made a group outing to the Science Museum, which is next door to Imperial in South Kensington. On our tour through the galleries, we stopped off in the optics section and there was a small interactive display where you could answer questions on lenses. It was basic GCSE stuff and I was getting them right. I was slightly stunned when everyone looked at me in surprise saying, 'Wow, you really know this stuff!' Up until this moment I think they considered me nice but dim, just as I had been seen as a child. But my studies had been very different from theirs, covering topics that they hadn't, so although my knowledge of their domain was limited, so was their knowledge of mine.

They saw that I knew what I was talking about, just not what *they* were talking about. Even those I was closest to in the group had thought I wasn't very bright, and it surprised me. From my perspective they didn't perceive me as equal,

CHAPTER 5

and this, once again, gave me something to prove. It was like a little burst of energy that bubbled up and reminded me to keep working hard.

The process of doing a PhD isn't always a pleasant one; it's supposed to be challenging, and I understood this and indeed thrived under the pressure. I was so used to pushing myself that, yet again, at several points during the four years, I was told by a few of my fellow students to stop working so hard as it was showing some of them up. Quite some feat for a PhD student. This has been a theme in my life. I look for opportunities and, if they come along, I throw myself into them, letting them consume me, making the most of things by prioritising and putting in long hours. They effectively become my hyperfixations.

The main thrust of my PhD was the further development of a piece of equipment that measured thin lubricant films in engines. We were using optics for these very fine measurements and could measure down to less than a billionth of a metre or a hundred thousandth of a human hair. The technique already existed but I helped refine it, building on two specialist techniques, spectroscopy and interferometry, and it eventually sold across the world. All our experiments were valuable, but this was one of the achievements I am proudest of, because of its legacy.

At some point in the PhD process, every student has to demonstrate that their work is up to snuff and unique enough to be considered for a potential PhD. This transfer process required the writing of a report as well as giving a presentation. Writing continued to be a challenge for me, but I blossomed in the presentation as I was so much happier talking about my work, and I passed, which was a relief.

At this stage, I was still fascinated by space, and although I no longer really thought I would become an astronaut, my ultimate dream was to be a space scientist. But I didn't tell anyone; I hid it in the background of my mind. I didn't think I was good enough to get there. I assumed that the industry was full of brilliant, bright people – literally rocket scientists – not people like me who always seemed to work harder than everyone else to achieve the same results.

I also thought, if I didn't speak of my desires aloud then nobody could disabuse me of them as they had done when I was a child.

Working in the lab, I was part of such a wonderful community. I got on well with Rob and his girlfriend, Julie, and also Gary, who moved to the USA and changed his name to Jim, so now we all call him Ga-jim. We would start to say 'Gary' and then realise our mistake and switch to 'Jim,' which resulted in a strange hybrid name that stuck. Howard was studying for his PhD while also running the family business, and David was in a similar position but he owned a property business. We were quite a mixed bag, to say the least.

There were a lot of rowers in the gang, partly because one of the lab founders, a Professor Cameron, had been big

CHAPTER 5

into rowing so encouraged people with the same interests to join. I became a sort of mascot, surrounded by all these big burly blokes. At lunchtime we would go to the senior common room, and they would clear their plates in an instant then start eyeing up my food, my main meal of the day, and asking if I was going to eat that sausage or finish my mash. I lost a lot more weight during this period but I know all of my food donations went to a worthy cause.

In the evening, we sometimes went out drinking, mainly at Belushi's, the bar at Imperial where we ordered pizzas, and I drank pints of cider (beer has always been too bitter for my sweet tooth). We sometimes had races to see who could down a pint the quickest; for some reason I was a really fast drinker, but it was a classic case of winning the battle but losing the war. After such a race I would be trollied, as my alcohol tolerance was not high. So I would stagger home and wake up the next day to cold pizza for breakfast. It was very good fun to be part of a tribe, but I was still lonely.

There were other single people in the lab, one or two of whom I would have liked to date, but I didn't know how to approach them. Like Hans who was kind, tall and Swedish, and took me out on his motorbike a couple of times. My feelings towards boys had always been complicated, tied up in my father's distrust of romance and anything distracting me from my studies. I didn't have very good role models with my parents either. My mother has been married four times. My father, married once, never appeared to have another relationship and remained angry with my mother until the day he died. There was an implication from my father that, if I wasn't careful, I would turn into my mother. This was something I was already aware of, not just from him, but from some of the choices she made around men.

When I was seven, I held hands with a boy in the school playground, but the rest of the children started teasing us and he backed off. From then I carried vivid memories of boys calling me 'Blackie' and telling me to go back to where I had come from, so I was very wary. Romantic relationships seemed like a quagmire that I didn't want to navigate. When I did have feelings for someone, it was always with the caveat that I could easily be rejected, mainly because of the colour of my skin.

I was and am proud to be Black; I love being Black. I have never wanted to be anything else. Some people looked down on that. I just assumed I wasn't valued, and this affected my view of boys, too; I thought they wouldn't value me either. Relationships had seemed so complicated and something to be feared. Shutting that part of me down had seemed like a good idea until I got to my early twenties and decided that my virginity was now getting in the way. I had to sort it out.

My first sexual encounter was a one-night stand. I was in a nightclub one evening. It was one of those times when I was happy with the way I looked and feeling good about my life. I got chatting to an attractive guy, which was fun, sharing information about our lives, and I told him I was

CHAPTER 5

a physicist. I went back to his flat, we had sex. We used a condom but it was a little awkward as I had no idea what I was doing. The next morning, over breakfast, he asked me more about my job as a fitness instructor. He had completely misheard me over the loud music. I took it as a massive compliment that I looked like I worked out all the time, but realised I was not the person that he thought I was.

He was a nice chap, but he was a little horrified when I told him I was a virgin. I knew we didn't have much in common and he probably felt the same because we didn't exchange numbers. I left, released from the burden of virginity and sort of glad that I had got it out of the way, but also sad, as I would have liked my first sexual encounter to have been more meaningful. What's more, it didn't stop me feeling lonely – in a way, it made it worse. I had been physically intimate with someone, without any kind of emotional relationship, and while it had been enjoyable on one level, it made me yearn even more for a deeper connection.

*

During the final year of my degree and the beginning of my PhD, my loneliness only grew. Sunday was the toughest day of all: estranged from my family and no longer a churchgoer, I would wander aimlessly around the local area. Father had still not forgiven me for seeking my independence. I even accepted a lift from a stranger once, just to have that human interaction. I knew it was a foolish to do and that things could have gone horribly wrong. Luckily, nothing did. It was one of the hardest times of my life. I don't think I have ever felt more insignificant.

Just before my PhD began, I decided to move out of Philbeach Gardens. I was still in contact with Gideon, who was now working on his geology PhD up in Leeds. He went home over the holidays, and I realised just how isolated I felt there knowing he was not in the building. I looked around at what was available. I found a bedsit in a terraced house in Beaufort Gardens, just down the road from Harrods.

As a child, one of the things that my father would do was drive my sisters and me around Harrods late at night. It was an odd outing for a family, especially at midnight, but I loved being in the heart of the city when all was quiet, looking at all of the shop's window displays, each with mannequins acting out various scenes. I loved the bright lights, the fancy clothes. It was especially fun around Christmas time when many of the displays would be animated. I would make up stories about each section. Little did I know that years later, Harrods would be my local shop, and I would breeze in to pick up a pint of milk.

The bedsit was tiny, a long narrow room with a single bed, a pull-up desk, a sink and a window at one end. It had a shared toilet and bathroom. With my love of long baths, I would quietly fill up the tub at 3am so I wouldn't disturb

CHAPTER 5

anyone or monopolise the facilities. I was the youngest person in the building and, when I bumped into my neighbours, we would always have a chat. It was really well situated, much closer to Imperial than the Christian Empire, and I was comfortable there.

Eventually I became reacquainted with my father. Our bond was a strong one and his anger finally subsided. This enabled me to visit him and Gracie on a more regular basis and help out when I could. I remember Father coming to visit me in the tiny Knightsbridge flat on one occasion. I was in contact with him to tell him I could not visit for a few days as I had food poisoning. Before I knew it he had arrived with some hot Nigerian pepper soup. I decided it was probably not the thing to eat with a bout of food poisoning, but the fact that he cared enough to bring some round made me feel much better in myself. Our relationship was not as it had once been, but the fact that we were now in regular contact was heartening. Still, there was always the fear that at some future date I could be ostracised again, which made me very aware of the security I had lost.

Life continued with this rhythm for me for a number of months until an interesting possibility popped up. Imperial had purchased Clayponds, a new-build complex in South Ealing for postgraduate and PhD students to live in and, after agonising over lists of the pros and cons of both places, lists that I still have today, I decided to take the gamble and apply for a room. Yes, I had been a glorious ten-minute walk from Imperial, and the journey from South Ealing would involve travelling in by Tube, but I would be part of a larger student community. I was desperate to interact with more people.

I was eventually allocated a bedroom in a two-bedroom flat with a kitchen and bathroom, but no sitting room. When I first arrived, I was there on my own and, because everything was brand new, I was unpacking all of the kit, from the brand-new fridge-freezer to the unused oven. Most student accommodation tends to be tired and well-worn, so it was a treat to be unpacking all this stuff. It was such a luxury after cooking on my little Baby Belling stove and using a tiny washing machine that drained into the sink. Suddenly I had a proper kitchen with all the amenities.

I had two more romantic encounters during this time, the first with a student at Imperial. We weren't openly dating but we had fun together and then, like a movie, I got together with my old uni friend, interrailing partner and bedsit mate, Gideon. We only became an item after he had left London, which meant maintaining a long-distance relationship. A different kind of loneliness.

We decided to go on a cycling trip around northern France in the summer break, filling our old interrail rucksacks with camping equipment so we could keep costs low, stopping off wherever we fancied. I was wobbly on a bike, to say the least. I had learned at the vicarage in Hastings, but I hadn't progressed on to roads, so I had to start again,

CHAPTER 5

as well as needing to build up stamina for many hours in the saddle.

I bought a brand-new steel bike, which was incredibly heavy, especially compared to the expensive carbon-fibre versions which were all the rage then. Nevertheless, it was strong and reliable. It also meant that it was not a desirable steal. When I first got it, I locked it up in one of the bike racks at Imperial and someone did try to saw through the chain, but they couldn't open the link and gave up.

I practised around Gunnersbury Park, near Clayponds. At first, I would have to get off the bike to push it up even the gentlest slopes, but as time passed and my fitness levels grew, I was able to cycle up hills without stopping. I thought France would be a doddle, but we spent a lot of time navigating busy roads in terrible wet weather. I was wearing a rucksack taller than me, pushing the heavy steel bike up steep hills, and we ended up only camping twice. We came home after ten days.

One of the benefits of this holiday baptism of fire was that it prepared me for anything on a bike, and I could now happily cycle around London, including the dreaded Hammersmith gyratory system which was the quickest way to get to Imperial. You couldn't go that fast because of all the traffic, so I felt fairly safe. Other than once putting my foot out for a kerb that wasn't actually there, I didn't have any accidents. I could cycle from South Ealing to Imperial in Kensington in about half an hour, which meant I could cut out travel costs and the long trudge to and from the Tube station in South Ealing to the flat via the cemetery. Not ideal for a lone woman. As well as getting incredibly fit, I loved the independence my bike gave me and, whatever time of night I was out pedalling, I didn't feel fearful. My bike was my trusted steed, my beloved companion – just like my father had had in Nigeria. It meant that when I visited family I was not worried about missing the last Tube home. I began to visit Hal on a regular basis, as well as Father and Gracie. No matter what the hour, I could leave and get home safely. Also, there was something wonderful about cycling though London in the wee small hours. All the places that were usually crowded became free of traffic: the midnight city, beautifully lit and completely accessible.

I missed Gideon when he returned to Leeds, and the distance between us was unsustainable. We drifted apart. I was also committed to establishing a social life at Clayponds, because that was one of the main reasons for moving there. It took a while to get going and I still felt isolated being in a flat rather than a shared house, but as more students moved in so the community grew, there were regular parties and I started going out more. My friends Sue and Baz invited me to go to one with them, which turned out to be Martin's 25th birthday celebration.

Martin was also doing a PhD in mechanical engineering, but he was in a different group from me, so we didn't really

CHAPTER 5

know each other. It wasn't exactly love at first sight. He had very short hair, which reminded me of men I had seen outside one of the local pubs where extreme right wing groups used to meet. Of course, this was the very opposite of who Martin was. I thought he was a bit nerdy, which is funny coming from a woman who began making her own telescope aged 14. I think he thought I was a bit odd.

We began meeting in tea breaks, a tradition in many university departments. There was a morning and afternoon coffee organised every day which we were all expected to attend. In some places it is compulsory. I didn't drink tea or coffee, but it was a great way to get different groups within departments to meet up and chat. To begin with it was quite alienating, as most people had a background in mechanical engineering and I was the only physicist. Martin had come from electrical engineering, so we gravitated towards each other, feeling like outsiders together for a while.

We started dating, which opened up a whole new world of people, parties and trips to the pub. Martin loved going out. He was surrounded by a great group of friends, not just from university but mates he had from primary school. The very opposite to me. For the first time, I had a busy social life, between my friends in the lab and Martin's gang, and I loved being immersed in it all. One night, a group of us from Clayponds entered the quiz at the local pub, and we won. The following week we returned, smugly saying to each other that we couldn't win every week because it

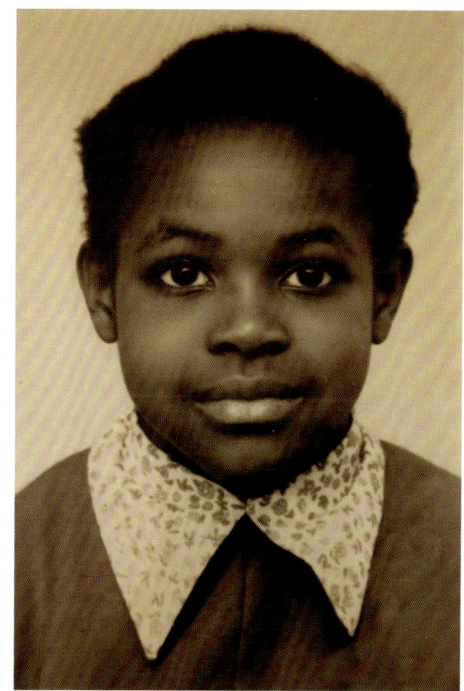

Top left: My father, circa 1980s.

Top right: Me around 7 years old.

Right: My sisters and I eating ice lollies. I'm the one in the pram.

Above: Me as a child.

Right: A photograph of me wearing one of many school uniforms, around 14 years old.

Below: We are family, I've got all my sisters with me!

Right: My glamorous mother with our pet rabbit during our years in Hastings.

Below: Reading up on the next destination on our interrailing trip, circa 1988.

Above: At a party during my PhD, circa 1991.

Left: Holding my baby, the bHROS (bench-mounted high resolution optical spectrograph) circa 2001.

Left: Getting fitted for my wedding dress, which I designed myself, in Bangkok.

Below: Lori, Martin and I at the summer solstice, Stonehenge, 2018.

Bottom: My mother, Lori and I – three generations celebrating my damehood at Windsor Castle, 2024.

Right: Lori and I meeting Tom Hanks, a fellow space lunatic.

Bottom left: Meeting Michael Caine at Microsoft's Future Decoded event in 2018, where I was his warmup act.

Bottom right: Jonathan Ross and I talking about will-o'-the-wisps in the Yorkshire Dales.

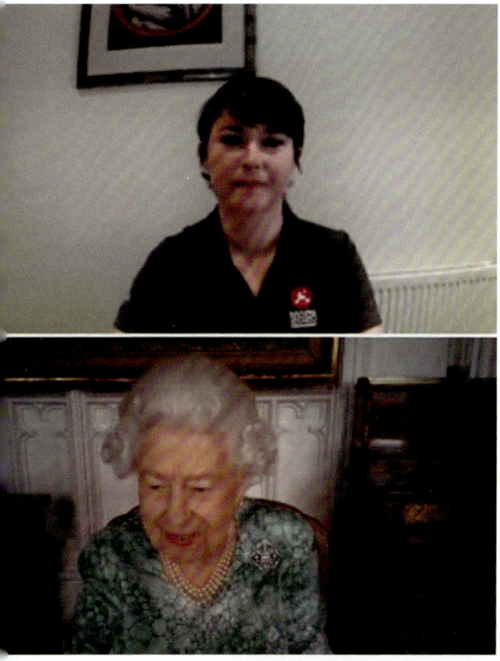

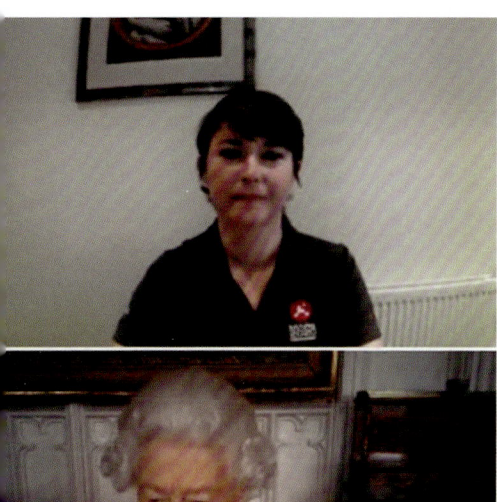

Top: A Zoom meeting with the Queen for British Science Week, 2021.

Bottom left: Myself, Tim Peake and Thomas Pesquet at the French Institute in South Kensington.

Bottom right and top overleaf: Me at the BAFTAs for the Television Craft Awards.

Below: Me at the Royal Institution for the announcement of the 200th Christmas Lectures, 2025.

wouldn't be fair – after all, we were clever PhD students. We were trounced! Justice was served.

After several months of living in the flat on my own, someone moved in without any warning. I knew it would happen at some point, but I thought I would be told in advance, but there I was, suddenly sharing a small space with a man, a complete stranger; to this day I am not sure I ever knew his name. I couldn't settle so I spent more and more time at Martin's shared house and got to know the people there.

I got a part-time job in the local video store. I could cycle there in ten minutes although I was often a few minutes late opening up on a Sunday after a late night of partying. Neil Kinnock, the MP, was one of our regulars, and I remember him waiting outside one morning.

We were allowed to play PG videos in the shop to keep the TV screens occupied. I would play a lot of *The Simpsons* in the store, which was my introduction to the show. I didn't have a TV or video player, but I needed to sort that out because taking videos home for free was a perk of the job.

It was a happy place to work, and I stayed there for a couple of years until the stress of my PhD ramped up. Even the robbery didn't spoil things too much. It was 10pm one night and my colleague and I had just locked the door and turned the closed sign round. The lights were still on while we cashed up and I spotted someone at the door, knocking

CHAPTER 5

to come in. This wasn't unusual, because we would often get people pitching up late to hand their videos back and avoid going over time thus incurring a penalty charge.

I unlocked the door, let the guy in and he whipped out a Stanley knife. 'Give me the money,' he threatened. We didn't argue. It wasn't worth being brave for £200, which is what he made off with. He was caught sometime later after targeting several shops in the area.

Although the actual event was quick and undramatic, I had nightmares for a while, dreaming of people jumping out at me, brandishing weapons. My biggest fear was that the police wouldn't believe me and think, as a Black person, that I was in on it. I had such paranoia about being accused of something I hadn't done.

*

The department for my PhD was mechanical engineering, but the group that I was in was called tribology, which stems from the Greek word *tribos* meaning to rub. To put it into simple terms, it is the study of friction, wear and lubrication. It helps us understand how surfaces interact with each other, like metal in engines, tyres on a road or bones on a joint. This can be abbreviated to lubrication engineering, which is what my PhD is in.

One night a group of us went to an improv comedy show and we each had to write down our occupation on a piece of paper, for the comedian to turn into a joke. I wrote 'lubrication engineer' – I think that was the easiest gag they ever had to work with!

The mechanical engineering environment was very male dominated, with a heavy hint of sexism – although not from anyone I hung out with – so being female singled me out. So too did being Black. I remember one of the academics said, 'Your forehead is darker than the rest of your face. Why is that?' I did not know how to respond.

People would assume I was the secretary of one of the professors or one of the support staff. We were all having lunch one day and a visiting lecturer asked me whose PA I was. There was a communal gasp at the table as my friends held their breath to see how I would respond. Although processing the situation, I was very calm, explaining that I was also doing my PhD, then someone changed the subject.

I have many of these stories. Years later, as a professional, I went to the offices of one of my contractors and walked into the reception wearing a suit and carrying a briefcase. The security guard handed me a bunch of keys and, when I looked confused, he said, 'the keys to the offices, love? So you can start cleaning them – start with the ones at the back.' The contractor was absolutely mortified when he heard what had happened, but the truth was that most

CHAPTER 5

Black women who came to that site were cleaners. Although I had hoped the suit and briefcase would have given the security guard a clue!

Then there was the time I was chairing a meeting as a new project manager. As everyone came into the room, they were bustling around finding a seat and one of them motioned to me and said, 'Coffee with two sugars, please.' I went and got them their drink and then I moved to the head of the table and said, 'Hi everyone, I am Dr Maggie Aderin and I am leading this meeting.' I think the person may have choked on their coffee. Yes, it was slightly wicked on my part – it really amused me – but I didn't want to belittle them or cause a scene. It seemed like the best way to make the point.

I am often asked how I deal with these situations. Firstly, there is nothing wrong with being a cleaner, a secretary or a tea lady. Being cross with someone for labelling me as such makes it seem like I think I am too good for that sort of role. Why is any job too menial for me? What I do object to is the assumption, when people see a Black person, especially a woman, and immediately jump to conclusions, but I don't take umbrage because being a cleaner is perfectly respectable.

I believe it is how you handle these conversations that counts. Do I respond angrily and shout, 'How dare you?' or do I try to re-educate people? I generally prefer to go for the latter. I have done a lot of re-education over the years. If I am furious with someone, they are more likely to be defensive and I am likely to be dismissed as an 'angry Black woman'. I want to talk to them in the hope that they don't make the same assumption again with someone else. For me, it is about breaking stereotypes and being listened to, which isn't going to happen if I get cross and shout. I am not going to start a fight. I want to respect the person I am talking to and engender a feeling of empathy between us. Though the occasions that this has happened to me have been relatively few and far between, if you have to face this sort of behaviour every day I can definitely see why the frustration would build.

I have often been underestimated because of the colour of my skin and my gender, but I don't mind as long as I have confidence in myself. Being an academic gave me kudos, and I felt I needed that. I guess I used my PhD as a rapier, the ultimate weapon to challenge the stereotypes I faced as a young Black woman. Come at me with your labels and I shall produce this qualification sword to show you just who I am. Swish! Take that! Now your stereotype is in tatters!

*

After a hardworking but happy few years, the end of my PhD was in sight, which meant writing my thesis: the part

CHAPTER 5

I was utterly dreading. I had climbed every other mountain, but what if this one was insurmountable? It was quite possible I would fail at this point. I am not sure about the percentages, but there are quite a few people who do the research needed for a PhD but never complete the write-up, and my biggest fear was that I would be one of them. My unknown dyslexia may have been waiting to trip me up and, without realising what I was doing, I found coping mechanisms which mostly meant immersing myself in my thesis for hours or days at a time, reassured by the slow but steady progress. In writing this book I have gained a better understanding of the challenges I face. I procrastinate and put the writing off, and it took me some time to understand why. I realised that I can read and write, not particularly accurately, but it is something that I can do. The real challenge is the amount of concentration required to do it. I need to fully immerse myself, and it is truly exhausting but worth it. So, while I procrastinate, I still do what I have to do, but there is always a physical and mental price to pay.

I didn't own a computer, but Martin had one in his bedroom, so I would be on it at night when he was asleep, partly so it was free for him during the day to write his own thesis, but also because it is easier to immerse myself when there are fewer distractions. I also liked working late at night, and I still do. To cover the irritating sound of tapping on the keyboard, I would play CDs on repeat while he slept. It turns out that this was actually more annoying than my typing. There are a couple of albums Martin says he can't ever listen to now because they were played so much while I was writing up. The fact that both of us were at the same point in our studies really helped keep me focused too.

My appalling spelling was a worry, and when auto-correct jumped in I wouldn't spot the discrepancies, like 'disk machine', which was part of my lab apparatus, being turned into 'dick machine'. Luckily my supervisor Philippa (who's still a friend) saved me from these perils and became my proofreader and human spellchecker, earning herself a big credit at the front of my thesis.

By a cruel twist of fate, I had my PhD viva and my driving test on the same day. It was terrible timing, but I had no control over that. I wasn't even sure I would be able to make the driving test in time because a viva can go on for as long as it needs to. A viva is the meeting where you have to present your thesis, discuss your research and defend your work with an external examiner. It can take anywhere from a few hours to days, although the latter is rare.

I was nervous as the viva can make or break your PhD, even though you should be fairly confident by this stage – your supervisor shouldn't put you up for something you aren't ready for. The fact that it involved talking a lot was the easiest part of the process for me. Everything went well. I skipped off to my driving test and passed that too. Two major milestones in one day.

Chapter 6
Getting to Work

About six months after I had finished writing up my PhD, I woke up with a start, a cold sweat on my forehead. I would ask myself, *What do you think you are doing just sleeping? You have to finish writing up!* The ghost of my thesis, and the gargantuan effort it took to complete it, hung around me for quite some time.

My father made an astute observation. He said that before my PhD, I would talk about it in a loud voice in public places letting everyone around me know that I was a PhD student. But once I had got it, these types of conversations would stop, and I would barely talk about it at all. He was quite right, and I worked out what was driving this odd behaviour. Before I achieved my PhD, I thought that there was a strong possibility that I would never make it to the end, so I had to make the most of the process and the fact that I was studying for it, as that might be as close as I got to the real thing. But when I made it against the odds, I did not need to talk about it because I had effectively achieved the crazy dream.

CHAPTER 6

One of my favourite films of all time is Steven Spielberg's *Close Encounters of the Third Kind*, an amazing film that came out at the same time as *Star Wars*. Two game-changing blockbuster sci-fi movies, both right up my street, but it was *Close Encounters* that caught both my heart and my imagination. It is all about a man, played by Richard Dreyfuss, who has an alien encounter and then becomes obsessed with meeting them again. This resonated with me very strongly. Not because I have had such an encounter, although I do believe that there is life out there in the universe, but because I understand that strong desire to get out into space. It's been with me all my life.

It has governed much of what I have done in my career. In the many talks that I give, I advise adults and children alike to have a big crazy dream, because by having one you can achieve so much more than you might think is possible. The dream helps us pick ourselves up when we fall down. It enables us to push harder and travel further, even when tired and fed up. You will achieve more just by striving, even if the dream itself remains elusive. Space has this done for me. But just like in *Close Encounters*, having such a passion is also scary. Too much single-mindedness can be overwhelming for oneself and those around one, but looking back, I am glad that it has always been there.

This driving force powered me through GCSEs, A levels, a degree and then a PhD, almost to the exclusion of everything else. But now that Dr Maggie had emerged from the system, what was to come next?

It wasn't an option for me to stay on to do postdoctoral work; although I had initially received industrial funding for my PhD, my supervisors Hugh and Philippa had put lab money aside to allow me to complete it. Some money was generated by me installing the system we developed for my PhD at industrial sites at organisations like Shell and BP, who were making engine oils.

But not doing a postdoc wasn't just about funding. By this time I had been at Imperial for seven years. I was ready to get out into the big wide world and earn my keep. Besides, while I was in the last months of my PhD I had been relentlessly applying for jobs, and I finally got lucky.

Initially I focused on the oil companies that I had been working with, and while it made sense to pursue a career with one of them, it was 1994, we were in the depths of a recession and people were being laid off everywhere. It was soul destroying to send out my CV, fill in an application form and either get a polite 'no' or no response at all. As part of some interview processes, I had to take a psychometric test, and the company would look at my results and point out my weak areas, predominantly spelling. No surprises there, but this was scuppering my chances. It was hugely demoralising.

CHAPTER 6

Eventually, I applied for a job at the science arm of the Ministry of Defence, and got it. However, to take the job I needed to be vetted and I had to sign the Official Secrets Act. This entailed not only investigating me, but also my immediate family. While I was in regular contact with Gracie and had reconnected with my father, my relationship with my older sisters and my mum was more complicated. During my estrangement from my father, Hal and I had reconnected. I always felt that Hal saw me. As I grew and developed I had become an adult and she recognised me as such. But I hadn't seen Mum for many years. And now I needed her signature on an official document. I thought this could be a good way to break the ice after so many years apart. I invited her to the lab where I was doing my PhD at Imperial College. This would be neutral ground, and gave me the opportunity to show her what I'd been up to.

She came with Derek, a lovely man she is now married to, and I had Martin with me. I think we both needed backup there for moral support. The atmosphere was strained, but civil. Both Mum and I were treading carefully: we were calm and measured. After being estranged for so long, it was also rather emotional, and it shifted the block in our relationship. I was relieved and hoped we would be able to rekindle the connection we had had when I was living with her in Hastings. We had got on so well during that time. With the benefit of age, I could see what a tough time she had been going through with her previous husband, and understood how she felt undermined by Gracie and me leaving. After our meeting, she signed the document and we maintained more regular contact.

I started my first job with the Defence Research Agency (DRA). I had some qualms about working with a defence organisation. As a child I was a pacifist, wanting to get rid of all wars, and the best way to do that seemed to be unilateral disarmament. Over the years my views had changed; I guess I became less of an idealist. Also, with this job, I could now stop sending my CV out and staring at my ever-growing overdraft.

The job was based in Farnborough, at the airfield, and my work was helping to develop a missile warning system, which did what it said on the tin: detecting missiles launched at an aircraft and alerting the pilot. In the past, pilots had to look out for missiles themselves and, if they detected one, they would launch countermeasures called flares, which were like big fireworks, burning hot and bright, designed to spook the missile and distract it from the aircraft.

To ensure pilots could focus solely on flying the aircraft, we were developing an automatic system to detect incoming missiles. To do this, we used different sensors to get an understanding of what a missile plume (which is the fiery tail at the end of the rocket) looked like. By looking at a plume through different parts of the light spectrum, such as visible, infared and ultraviolet light, we could create a profile that would allow our systems to detect the launched missile.

CHAPTER 6

We were looking at these three different wavebands to reduce the false alarms: if a reading came up positive in all three, we could be confident that it was indeed a missile rather than reflection of sunlight from a passenger plane or metal surface. It is all about understanding the energy profile.

Other colleagues in the team were looking at countermeasures, the pyrotechnics that could spook the missile and get it to deviate from the aircraft. As we developed the countermeasures, the missiles grew more sophisticated, and so the countermeasures had to get better to keep up.

We went up to Scotland where they tested missiles and we flew equipment on Tornado fighter jets to get images of them. We also travelled to Pendine Sands in Wales, where they had a track which was one and a half kilometres long. On this track we could place a missile (with no warhead), light it up and look at the output signature of the plumes. We would fly above the missile in a small Andover aircraft, capturing the data on three different cameras to capture the three different wavebands.

There is a photograph of me, standing in the open door of an aircraft in flight, with a big belt around my waist, looking down at the track as a missile flew past. I like to show this picture to the students that attend my talks, and I say, 'This was my first job, hanging out of the back door of an aircraft. Like James Bond or Lara Croft.' But I do also add some small print, which of course needs to be read very

GETTING TO WORK

fast: 'Jobs in science can vary; if you become a scientist there is no guarantee that you will get to hang out of an aircraft.'

Less movie-star glamorous was the office in which we did the day-to-day work. It was a Portakabin just off the airfield that I shared with my boss. My boss's boss had his office in one of the main buildings. The main thing I remember of my few encounters with him was the array of posters of scantily clad women on his walls. Although it was just 20 years ago, these were very different times.

I came in at the rank of higher scientific officer because of my PhD. Even though I worked in the science division of the MOD, our ranking was directly connected to the military infrastructure, which is very hierarchical. My boss was the senior scientific officer and, when I started, I think he wanted to put me firmly in my place. Maybe he thought I was going to be a bit full of myself, coming from Imperial College with my PhD. So the first thing he had me do, on my very first day, was put some blinds up in his boss's office. I don't know if this was a test to see my reaction, but I really liked DIY so I enjoyed this challenge – which probably meant I passed. I spent the rest of the day looking at paperwork to familiarise myself with the project. I definitely preferred hanging the blinds.

To begin with I found my job a bit boring: lots of reading and not much hands-on work, but then I was tasked with running the tests at Pendine Sands so I had to gather all the lab equipment we were using and get it on board

CHAPTER 6

the aeroplane. Finally, this was something I could get my teeth into. The aircraft had been the Queen's Andover, but was now defunct and resided at Boscombe Down airbase on Salisbury Plain. Although I was based at Farnborough, I drove down every day and worked with the crew there to get everything fitted into the plane. They were a great team to work with and I was my own boss, doing my own thing, and I loved it.

I used to go to Shrivenham, a military college which also encompasses training in strategy and scientific knowledge. As a higher scientific officer, I was housed in the officer's mess and needed to learn all the trappings which were associated with being there, including the appropriate outfits for dinner. On one of the training courses, I had to get into a tank. I tried to climb into the tower of the tank and as I slipped in, my skirt went up over my head, so I arrived pants first. The first thing the male soldiers already in the tank saw of me was my knickers. I made light of it and apologised breezily for flashing them, but inside I was embarrassed; it was not my best entrance.

By this point, having completed our PhDs, Martin and I had to move out of Clayponds and we decided to rent somewhere together, our first home away from family and university life. We found a little bungalow called the Old Surgery (it had been the old doctor's), at the bottom of a steep slope that was very close to the railway line in Brentford. We could hear the freight trains at night. The house was lovely; it had been derelict for years but had now been nicely done up and we were the first people to live in it.

The problem was that, after years of being exposed to the elements, it was damp deep down in its very bones and, unbeknownst to us, the damp continued to come through the freshly plastered walls. The first sign of trouble we picked up on was when my asthma and eczema started getting really bad. Then, one day, I was getting some clothes out of the back of the built-in wardrobe, only to find they were covered in a thick furry mould. We had discovered the culprit. We contacted the landlord and he arranged for a dehumidifier to be installed, which helped a bit but did not get rid of the fundamental problem.

One of Martin's requirements was that he wanted a garage for his car, a Mark II Ford Escort. The Old Surgery had one, and a large outside space around the back with enough parking for at least three cars. Beyond the parking was an area of garden we called the wilderness. We found raspberries, gooseberries, redcurrants and blackberries back there, and all sorts of other things just growing wild.

It was an era of firsts; once I had a job and we had our home, I bought a car. Just like my bicycle, this was a real game-changer for me, because it meant I didn't have to commute by train to work in Farnborough, a convoluted trip involving several changes. The day after I passed my driving test and got my PhD, I decided I was going to drive to Farnborough, which meant taking the motorway. Martin

CHAPTER 6

wisely advised against this and thought I should build up to it, but I was keen to give it a go.

My car was a Ford Escort Estate in a nice rusty orange colour. I was so proud of it and polished it with T-Cut, a polish that takes off the top layer of dull, oxidised paint and reveals the shiny car underneath. It had a small engine and only four gears, and it was a bit lacklustre in the horsepower department; I had to work hard to build up speed, but that was no bad thing in those early weeks of driving to work and back. Again, I just loved the independence that the car gave me. The car felt like an extension of me, and I got a little cocky with it. Even now, I have a whiff of the petrolhead about me, and I will go out and about in my car (now electric) to relieve stress, pootling around local villages.

I was given a hire car to drive to Pendine Sands where we were working at the missile test range. One weekend, Martin came to visit me and we drove down to the beach. We noticed people were driving across the sand, which looked like a lot of fun. Several historical land-speed records had been set in Pendine, due to its vast expanse of open sand. We couldn't resist so I joined in, scooting around and picking up speed. I find this sort of thing exhilarating, particularly going fast. What I hadn't realised in this situation was that I had to stick to the hard wet sand and, as I gave an obstacle a wide berth, I got caught in the soft sand and my wheels started spinning. We were stuck on the beach in a car, which was not my own, and the tide was coming

GETTING TO WORK

in fast. I tried everything to get the car to shift – changing gears, revving the engine, reversing – but it only sank further into the sand. As the tide came in I started to get really worried. I was convinced that the car was not insured off the road, and it was going to be a hard one to explain to my boss. I even thought about calling someone from the military base, which was very near by. It seemed like a better idea than letting the car get flooded by sea water.

Amazingly, a group of guys in another car realised what was happening and pulled up next to me. They whipped out some boards they had in their car, dug around us and stuck the board under my wheels, which helped me to reverse out safely. They did all this in minutes, barely saying a word. They were well-versed in beach driving and getting stuck, which is why they had the boards ready. Another moment in my life where I relied on the kindness of strangers.

During my time in the job, I was also tasked with driving a 7.5-ton truck full of important equipment from Farnborough to Salisbury Plain. I had never driven anything bigger than a car, but I wasn't about to let that stop me. Surprisingly, a 7.5-ton truck is the largest thing that you can drive with a standard licence so I thought I should be capable of doing this, but I was advised to drive around the airfield a couple of times to get used to such a big vehicle before I hit the open road. Eventually I set off on my journey to the Larkhill military base, my heart in my mouth the whole time. I arrived in the car park after the 50-mile journey in

CHAPTER 6

utter relief that nothing had gone wrong, manoeuvring the truck into a parking space. But when I hopped out of the elevated cab I noticed several people in the car park laughing at me. It turned out I had parked too close to a thin lamppost, caught it and bent it out of shape. I had done the entire journey without mishap, only to ruin the dismount.

Later in my career, I was faced with a seven-ton nemesis again. I was at UCL and in contact with the UK Astronomy Technology Centre (ATC) in Edinburgh. They were lovely people to work with. We would collaborate to move components for some of the instruments we were building, and occasionally I was able to utilise their software and modify it for our equipment. When someone in the Edinburgh team mentioned they were getting rid of some old video conferencing kit, I jumped at the chance to take it off their hands.

I whizzed up to Edinburgh on the train with a colleague of mine and went to collect the transit van I had booked so we could drive the kit back to our laboratory at University College London (UCL) on Gower Street, just around the corner from Tottenham Court Road. When we got there, though, they apologetically explained that no transits were available, and instead gave us the keys to a 7.5-ton truck. Look, the half is very important when you would prefer to be driving no tons.

To add further angst to my situation, my colleague Andy suddenly realised he had forgotten to bring his driving licence, so he couldn't be signed up as the second driver.

GETTING TO WORK

There was nobody to share the driving with; I had to do the whole thing on my own. My first journey in such a vehicle had been a mere 50 miles. The journey from Edinburgh to Central London was just over 400 miles. As project manager there was nothing I could do but make the journey with my fingers gripping the steering wheel, shoulders up about my ears and buttocks clenched for eight hours. Every time I touched the airbrakes there was a hissing sound, and I took it personally.

*

After three years at the DRA – one of only two women in a department of 120, and the only Black person – I was ready to take the next step to senior scientific officer, so I applied for a role working in landmine detection at the Defence Evaluation and Research Agency (DERA), the new incarnation of DRA, and was successful. The United Nations has estimated that at the current rate of detection it will take over 1,000 years to clear all of the deployed landmines globally. Not only do they cause physical harm by maiming people, but huge areas of land can't be used because of their presence, and they leak toxins into the soil. I was focused on anti-personnel mines, some of which were hard to find through metal detection because there was very little metal in them. They looked just like plastic toys, so

CHAPTER 6

children would pick them up and they would detonate, causing terrible injuries.

In my new position, I was managing a team of six people, and we were focused on finding an effective way to detect landmines. The first step was to sweep the ground for anything with metal content. Of course, there are many metal things that can be found on a battlefield, from old ordinance to objects as basic as ring pulls from fizzy drink cans. All these things give a false alarm, and each alarm needs to be investigated carefully. To reduce the false-alarm rate we used a method called ground-penetrating radar (GPR). The GPR instrument sends a beam of radio waves down towards the ground. Most of the radio waves are reflected off the surface, but some penetrate the ground and reflect off objects below the surface. The reflected radio waves can be interpreted as an image, which gives an indication of what lies below. GPR systems are used on archaeological sites to find ruins, and also by the police to find buried human remains. Having identified a likely landmine candidate using these first two methods, the final sweep was done using a method called nuclear quadruple resonance (NQR).

The NQR system, like the radar system, sends out radio waves. But in this case the radio wave is of a very particular frequency, a frequency that is absorbed by the nitrogen atoms that make up the explosive's molecules. While there is nitrogen present in the ground in other matter such as fertiliser and decaying leaves, this particular frequency only excites nitrogen within the explosive molecule. These radio waves don't trigger a mine to explode; the energy is just absorbed by the atom and then rereleased as radio waves. So, if these specific radio waves are detected, it means that there is likely to be the presence of the explosive, and the object is likely to be a landmine. At this point the area of land would be quartered off, and the next section would go through the same process.

This three-pronged approach helped eliminate the majority of false alarms, but it was a slow, laborious process. Still, it's better than probing the ground with a stick.

The work in this field was very rewarding as it felt as if we were making a real difference in an area that posed a genuine global problem. Unfortunately, although the work was fascinating and I was glad that I got the promotion, I had to navigate a rather nasty boss who was quite a bully and reduced me to tears a number of times (although I never cried in front of him). I felt vulnerable and hated the environment, but I loved the work, so I stuck at it. I travelled and spent time with the Royal Engineers, listening to their experiences of landmine detection out in the field and incorporating their first-hand knowledge into the project. This was a couple of years after the death of Diana, Princess of Wales, who had raised awareness around a campaign to ban landmines. Her involvement made a significant impact for the cause, leading to a global ban. Sadly, she was not alive to witness this, but her legacy lives on.

CHAPTER 6

It was around this time that I made my first television appearance. I had been collaborating with a chap named John Adams, who made metal detectors. John had been interviewed by *Tomorrow's World* in the past and they had returned to catch up with him. When he told them about the technology we were working on, they asked if they could visit me in my lab. The show had always been a big part of my childhood and Maggie Philbin, who had presented it in the 1980s, was one of my heroes, so there was no hesitation. I talked about our achievements in landmine detection. The head of DERA came to see me and said how impressed he was and keen for me to take on more public engagement work. Who wouldn't want a Black female scientist with the gift of the gab?! It planted a seed in my mind. I really enjoyed talking about my work and sharing my knowledge in an accessible and interesting way. I wanted to do more of it.

Chapter 7
Father, Brother, Teacher, Philosopher, Friend

It was 1999 and, as we raced towards a new millennium, I wanted a new challenge, something that would take me nearer to my dream scenario of working in the space and astronomy field. I had a couple of false starts, including applying for a position to work on a telescope in La Palma in the Canary Islands. It was such a lovely opportunity, and I really wanted it, but it was in the electronics department, which was not my strength. I wasn't successful, but they said they would be in touch when they had something in optics.

Martin and I were living in an unfurnished maisonette in Isleworth, filled with furniture we had been given by friends and relatives, and we bought our first washing machine (albeit secondhand) which felt like a terribly adult thing to do. We were under the Heathrow flight path, and when Concorde flew over, everything in the house would rattle and we couldn't hear each other speak! We got so used to it that when friends came round, they would look at us in shock and hang on to the furniture, but we barely noticed it.

CHAPTER 7

I looked out of the window once and saw Concorde flying in formation with the Red Arrows. It was in celebration of the 50th anniversary of Heathrow airport, and we had a ringside view of the celebrations.

It was funny because, when my sisters visited us in Isleworth, which is on the Tube, they thought we lived in the sticks, because it felt so far out of London and had a sort of rural air.

It was around this time that a new opportunity arose. I got a job at UCL as a project manager, working on an instrument for one of the largest telescopes in the world. The lab was based at the Mullard Space Science Laboratory (MSSL) in Surrey, although I would have to spend a lot of time in the basement at UCL in Central London. My background in telescope making in my teens and my management experience in a military environment seemed to be the perfect combination. It was the closest to my space dream I had ever been.

The plan was to make an instrument to sit on the Gemini telescope, one of the most advanced telescopes in the world, situated on a mountain in Chile, but the project was having a few issues.

We were building an instrument called a high-resolution optical spectrograph (HROS), which is designed to analyse light from distant objects. A spectrograph is the workhorse for all large telescopes, as it allows you to analyse the light emitted from a wide range of cosmological bodies.

FATHER, BROTHER, TEACHER, PHILOSOPHER, FRIEND

The analysis involves taking the light gathered by the large mirror of the telescope and stretching it into its component colours using prisms and other optical systems, effectively creating a huge two-dimensional rainbow in the lab called a spectra. With this technique it is possible to tell various things about the object being observed, for instance if it is moving towards us or away from us in space, known as its red or blue shift. It allows us to do remote chemical analysis, identifying the elements and molecules in the atmospheres of galaxies, stars, planets and moons. As well as chemical analysis, we can measure temperatures, densities and, in the case of gases, pressure. The real joy is that all of this information can be gathered for objects light years away from us.[5] So, using this technique, we can work out what is happening in the heart of a star or the centre of a galaxy.

Our client, the telescope manager, was a wonderful chap called Doug Simons, who was based in Hawaii. Communication involved lots of late-night phone calls for me and early morning chats for him. It wasn't until I started the job that I found out that there had been eight project managers before me. This didn't bode well, but I wasn't perturbed.

---

5. As an aside, a light year is a distance that equates to the distance light travels in a year. As the speed of light is 300 million metres per second, a light year works out as a whopping 9.5 trillion kilometres (or 5.9 trillion miles).

CHAPTER 7

In my brief handover with the previous manager, he told me not to let the team slack. He wasn't inferring that they were lazy, but he had set deadlines that he wanted me to uphold with them. This was an interesting stance. The first thing I did was to individually speak to every member of the team informally over coffee, to find out how they felt the project was going and what they believed we should do to move it forward. There was a provisional design review (PDR) coming up, the first of two key design reviews which all space or ground-based instrumentation need to go through and pass – and nobody felt ready for it. The previous manager had wanted to forge ahead to keep momentum going.

After talking to my team, though, I decided to delay the PDR for six months. It was clear that we weren't prepared for it, or in the right mindset. I think that they felt no one was listening to them, especially after so many project managers. We were able to use my joining the team as an opportunity to assess the progress that had been made so far and to work out what was really needed to achieve a successful PDR. Most of all, I think that the team felt as if their expert knowledge of the project was being acknowledged and they were being listened to. When we did go for a review, it went very well, so I am glad I went with my gut. It also proved something to me that I had been aware of on previous jobs: together, we could achieve the seemingly impossible. This is a mantra I repeat today, even as someone who values her independence. We can achieve so much more if we join forces with others.

We had just passed the PDR, but we still had a way to go to pass our second key review, the critical design review (CDR), which gives you licence to start cutting metal to manufacture the instrument. The instrumentation we were building was very complex and we wanted it to be attached to the telescope. This meant that every time the telescope moved, pointing at different parts of the night sky, the instrument would flex and shift around. We were trying to find a system that would counteract the flexing, but its implementation was proving to be very challenging.

Eighteen months later, we went through the CDR. This time it didn't go as well. The light from the distant object, gathered by telescope, was being fed through the HROS, but as it bounced off various components, spurious images were created which could overwhelm the real data. The movement of the telescope was as expected, causing the instrument to flex, so we had a laser going in the opposite direction to monitor the movement and take it into account. The problem was, it wasn't working. We didn't pass, which could have meant project closure, but thankfully an alternative plan was put in place. We had around a week to redesign the whole system, so the team worked together, burning the midnight oil. I remember that as we came to a close I pulled one too many all-nighters and fell asleep standing up on the crowded commuter train going home;

CHAPTER 7

I had to be caught by other passengers. The solution was, instead of attaching the instrument to the telescope itself, we placed it on what is called an optical bench, in the basement of the telescope. The optical bench is a stable platform which all the optics could be mounted on. This meant that they were no longer attached to the telescope, resolving the flexing problem. To get the light gathered by the telescope to the instrument, a fibre optic cable was used. This was a compromise, as some light is lost when passed through a fibre, but it was a trade-off between stability of the system vs. its efficiency. HROS was literally benched. We called the phoenix that emerged from the fire bHROS, the bench-mounted high resolution optical spectrograph and, when it passed the next CDR, we began construction.

*

While I was working strenuously on bHROS to keep the project going, my father's health was sadly deteriorating. Gracie was studying for her degree at university, and Father was on disability benefits, which meant money was tight for them, so I gave some extra financial support. I did this willingly but was also niggled by the fact that Father insisted on giving the church he belonged to a tenth of any money he had. He was getting no support from the church, no home visits and no calls as far as I could tell, and the pastor was driving around in a Rolls-Royce and maybe even had a private jet. My father was struggling financially, and I was supporting him, which I was very happy to do, but I was not happy for a proportion of that to be given away. I understood that this was my father's way of expressing hope. The hope of something better in the future. That giving this money was hopefully ensuring a better life for himself. But I did feel resentful of the pastor who could have given him a lot more support.

Father and Gracie were still living in the council flat where I had grown up, but with his growing infirmities it was almost impossible for Father to move around. There was no lift in the building where he lived so he was now housebound, and this exacerbated his feelings of isolation and the walls closing in on him.

I contacted the council, who gave us two options. The first was to try and find someone to house swap with; the second was that they would attempt to find him a place, but that was likely to take a lot longer. We investigated a house swap, although the flat was in a bad state. Eventually the council found a flat in Holborn with a lift, and I was able to transfer father and Gracie there. I also got him a wheelchair, which meant I could take him out and we would make a detour to a favourite stationery shop – another interest we shared.

A skin graft was recommended for his leg ulcers, which would have given him a semblance of normal life, but he

CHAPTER 7

didn't trust the doctors and resisted hospitals. Instead, nurses would come to change his dressings twice a week. I was there when they were doing it once, and his cries of pain as they changed the dressing still haunt me today. It seemed crazy to me that there was a viable solution, and even the nurses agreed that there was no choice but to have the operation. His open wounds were at a high risk of infection, and medical intervention could give him back a better quality of life.

I took the decision that he had to go into hospital. I weighed up the doctor's advice, the state my father was in and all the potential outcomes, and I ignored his protestations. I say 'I', because there was nobody else to make the call. My father had fallen out with my two elder sisters some years before, so they were not involved in his life, and Gracie looked to me in situations like this.

I went with him to the hospital and regretted my decision as soon as we arrived. The place was overcrowded and there was no space, so he was left on a makeshift bed in the corridor and, as the day turned into night, I was sent away. I had to leave my father, blind and panicked, with no idea of what was going on around him, and the worst thing was that I had imposed this on him.

By the time I returned the following day, they had found him a bed on a ward, and he stayed in hospital for several weeks. I visited him every night, bringing food, talking to him and the nurses, checking he understood what was going on. One of the challenges of being visually impaired is that you are unable to recognise the staff that you are interacting with. One day I turned up and the ward Father was in was empty; he told me everyone else had caught MRSA and been moved, but it appeared he had escaped it.

The skin graft was done, but it wasn't a huge success. Maybe it was too late and, the ulcers were too bad, but it never took. We also discovered that he actually *did* get MRSA, which had most likely weakened his immune system. As a result, he came out of hospital perhaps worse than he went in. And I felt totally responsible.

Father settled back at home, and I would go round and check on him, collect medication and bring him shopping. One night he called me, saying he wasn't feeling great. He wanted me to get an over-the-counter antacid medicine he used to take, because he was having chest pains which he thought was acid build-up. He asked if I could go to the chemist on my way to work the following morning to collect it and drop it off to him.

The next morning, I came into London earlier than usual so I could go to the chemist before work at UCL, which was fortunately just a few minutes away from my father's flat. It was a clear crisp September morning and I remember thinking that I didn't need to rush. This is my overriding memory of that day. I dawdled. The chemist didn't have the medicine he was after, so I went straight round to the flat.

There was no answer, but I had a key and I let myself in. Gracie wasn't there, nor was my father, and I couldn't get

CHAPTER 7

hold of them on the phone. Something didn't feel right. And then Gracie called me back. Father had been taken away in an ambulance, she said, and she thought he might be dead. She didn't know what to do. He hadn't been responsive so she had splashed water on his face to rouse him before calling the ambulance. She felt she should have called them earlier. I told her that she had done the right thing, and that it would have made no difference. I felt guilty too for taking my time that morning. Would things have been different if I had got there just a little earlier?

I ran outside, wild and crying, and saw a taxi parked up. The taxi driver said, 'Sorry love, I'm off duty.' I said, 'You don't understand, I have to get to the hospital immediately. My father might have died.' And he said, 'Hop in, I'll get you there as quick as I can. No charge.' You see, the kindness of strangers again. Another example I keep in my pocket of good people doing good deeds.

At the hospital, Gracie and I were given the news we were dreading, Father was dead following a suspected heart attack. It was 2001, two days before 9/11. When he passed away it felt like my world was coming to an end and then, a couple of days later, it felt like the whole world was ending. Nothing made sense. Soon, though, the project manager in me kicked in. There was a lot to organise, people to contact and a funeral to arrange; this was my salvation in those early days. I put the emotions to one side and made the first calls to Sue and Hal, and then my auntie, my father's sister.

FATHER, BROTHER, TEACHER, PHILOSOPHER, FRIEND

My auntie wanted to take Father's body back to Nigeria, which was a nice idea, but he hadn't been there for over 30 years and I wasn't sure that would have been his choice. She also refused to agree to a postmortem. This was required by law because the circumstances of his death were unclear and there was no recent medical intervention. He could not be buried until this happened, but my aunt would not agree to it. While this played out Father was, for want of a better word, in storage, until it was clear to my aunt that there would be no funeral without an autopsy. I had no idea what was involved in the aftermath of a death: finding a plot, sorting out the death certificate. A whole host of things which, at the age of 33, I had never encountered before.

The two biggest challenges were how to celebrate Father's life and where to lay him to rest. I was convinced he would want to be buried in consecrated ground and found a plot, but my elder sister overruled this, saying it wasn't a nice spot. She had found somewhere quieter and leafier, but it wasn't consecrated. It haunted me during the days leading up to the funeral as I was preparing the ceremony, and I rang various Church of England vicars in the area of the cemetery to ask if they could consecrate the ground, but there was no joy.

Organising the funeral was hard because Father had not discussed what he wanted. I found somewhere to hold the wake and sorted out some simple refreshments. I did everything myself because my father had been estranged from

CHAPTER 7

most of the people in his life and I didn't think he would want others to take control, but I can see how this might have made my elder sisters feel pushed out. Gracie's input was key because she had been living with Father and would hopefully know his wishes best. So between us we tried to think about what he would have wanted, but I was worried we might get it wrong. These were effectively his last wishes, so it really mattered.

For the ceremony I listened to various pieces of music, trying to find something that would capture our loss, and in the end I settled on Purcell's aria 'Dido's Lament' with the powerful lyrics of 'remember me'; it still pulls at my heart when I hear the opening bars. It felt like the only thing I could give my father now: the fact that he could live on if we remembered him.

I planned the ceremony, booked the chapel, invited relatives and made the orders of service. Martin helped me with the latter. In my sleep-deprived state I was finding it hard to work out how to print the pages to make a small order of service booklet. Even finding a picture of Father to put on the cover was hard, and in the end I used an old passport photograph of him from when he was happy and still able to see.

On the morning of the funeral I took the train into London with Martin. I was effectively a zombie due to lack of sleep, spending long hours organising the funeral and suppressing my grief. I was still on the phone because I had not secured the consecration of the ground that Father was due to be buried in in just a few hours.

But I managed it, I found a vicar who was willing to come that morning and perform the blessing. So, before everyone else arrived, Martin and I met him and he kindly did the consecration of the land where Father was to be buried. I was so relieved that I had been able to pull this off. I knew how important it would be to Father and, in that moment, I wasn't project-managing a funeral, I was a daughter in mourning; I howled at my loss. The consecration was my last act of love to the man who had guided my life and given me so much.

The funeral itself was a very rare gathering of most of the family, including my very vocal auntie. I was back in logistics mode until the reception, when I was overwhelmed briefly and cried on my mum's shoulder, something I hadn't done since I was 14. She was lovely and it was reassuring to feel that I still had a parent, but my father's absence was still yet to sink in. A few weeks later, I woke up one morning and started sobbing as if I would never stop; this was when the reality of the loss of my father properly hit me.

Gracie was also felled by it. First I took on the challenge of making sure she could stay in the flat. But, once this was resolved, she suffered a series of seizures, ending up in hospital. Eventually my sister was diagnosed with epilepsy, given medication and sent home. She was in her late twenties by then and living on her own for the first time in her life.

CHAPTER 7

I spent a long time contemplating what to write on Father's gravestone and how to reflect this complex man. I thought of what he represented to me and my sisters, and I settled on 'Father, brother, teacher, philosopher, friend'. He was all those things to me, particularly a philosopher, because he believed in thinking for yourself and he encouraged that ability in each of us. I can't think of a better gift to give a child.

*

Loss affects everyone differently. I once received an email from someone who had read one of my first publications, *The Book of the Moon*. It is a celebration of all things lunar, and I like to think of is as my love letter to the moon. In the email, they said that they used to moongaze with a relative of theirs, but unfortunately this person had died and now they couldn't look at the moon any more without regret. They felt haunted by it. I felt compelled to reply as it resonated so strongly and made me think of my father. The moon is my lodestone, it calms me when I see it, so the loss of the moon seemed awful to me. I emailed back telling them they were looking at it the wrong way round. They should see the moon as a conduit to keep the connection with their loved one, a positive, happy reminder of the time

FATHER, BROTHER, TEACHER, PHILOSOPHER, FRIEND

they spent together, and to celebrate them. I was writing to this stranger, but I was also speaking to myself. The moon always reminds me of my father and the stuff we enjoyed doing together. All I have to do is look up at the night sky.

Chapter 8
Milestones

This is where one part of my life ends, and another begins. My father had been my constant, a safe place and an anchor throughout my life, for better and worse. I was not blinkered to his flaws. He encouraged a curiosity and drive that I have relied upon throughout my life, and made a career out of. For everything he may have unwittingly taken away, he gave me so much more. From this point on, he would not be there in my life, and I was bereft.

As my work on bHROS progressed, Martin and I left London and moved to Woking, renting a bungalow next door to a pub, which was very useful when we got home late and didn't want to cook. This was the first time I had lived outside the city since I was a child, with no Tube to jump on. Martin and I had a running joke that every time we moved house, we went slightly further west – even if it was just a mile, we always seemed to be going in one direction. As he is from Somerset, it felt like we were inching back to the West Country, doing it slowly so I wouldn't notice.

CHAPTER 8

I spent much of my time at UCL, but every so often I would work from the MSSL, based in a stately home just outside Dorking. It was named after a rich entrepreneur who had donated the building to UCL. It has the most glorious grounds, and I loved arriving early, after dropping Martin off at work, to see mist on the lawn and deer bounding into the trees. Our meetings were held in the wood-panelled library, the opposite of the clinically clean, white laboratories where we were building instruments for space. (We often film *The Sky at Night* there, so I return regularly.)

If I was working at UCL in Central London, I would be on the 7.32am train. That time is still burned into my mind. I was an early adopter of working on a laptop on the train, although my first one was the size of a small elephant and weighed about as much. I was less bothered about having a mobile phone as, when I did, that just added to the heaviness of my bag. Less businesswoman, more packhorse.

The project surfed success and failure over a number of iterations, and this brought its own challenges with good people coming and going. One guy said there was no incentive to work more efficiently because it would mean he'd be out of a job sooner, an attitude that was hard to know how to handle. For the most part, though, it was a strong team with a few familiar faces from my MOD days who applied for positions I advertised and joined me at UCL.

We had optical people focused on the optical system designing components that would create the spectral/

rainbow, and mechanical people designing structures to hold those optics in the precise location they needed to be in. Electronics people would develop the mechanics to allow it all to move, and software people created programmes that moved the mechanics. I project-managed a diverse team of around 20 people in the end, all working together. That doesn't include the people in the workshops manufacturing it; it was a big group of experts.

The huge slabs of glass we used must have weighed around 20 kilos, and the prisms around the same. They had to be really high-quality glass, so they were made in Holland or Germany, and we would check the homogeneity – the quality of the glass. Any inconsistencies in the density could give false results. The challenge didn't stop there. We had to send the high precision glass to other people to shape into the prisms, mirrors and lenses we needed. One of the companies we used was based in caves in Chislehurst. The small, dank workshops there were ideal places for glass shaping. We could do a significant amount of steelwork and support-mechanism construction at UCL, but other elements, like the glass, had to be contracted out.

Because the spectra was very sensitive to thermal variation, the instrument also had to be kept in a temperature-stable environment to stop small fluctuations in heat which could affect the results. But the detectors that captured the rainbow had to be cooled so they could work at maximum efficiency. As the project manager, I looked after deliveries,

CHAPTER 8

timescales and budgets, but I was so fascinated by the science that I eventually took on the role of systems engineer too. This meant coordinating all areas of expertise to bring the whole system together. I loved it. I learned so much and often had to step in if we lost someone as the project progressed, while outsourcing areas if we did not have the expertise in-house. The ATC in Edinburgh came to our rescue on more than one occasion. As many of our mechanisms were similar to instrumentation they were building, we could adapt and use much of the software they had generated.

I enjoy being part of a team, even if I am something of a lone operator. If I can do things myself, I will, but I also know when it's not possible. I took on as much as I could appropriately handle. Yes, I was learning on the job, and that meant being thrown in at the deep end several times, but I also recognised when I needed to hand something over to one of the team or employ a private contractor.

Towards the end of the project, some of our funding stream was slow so project bills weren't being paid. Progress was dragging so, to keep the job going, I started to pay for things myself with my personal account. Madness, I know. It was a dangerous practice which drained my savings. By the time the project was finished, I was owed around £10,000. There was a moment when I thought, *What happens if they don't reimburse me?* I wouldn't recommend sinking your earnings into your work, but without doing so we may not have delivered the project, and we did – which was a joyous result! And yes, I did get paid back.

One of the highlights for me was the travel associated with the project. I went to Hawaii where the Gemini South's twin telescope was set up. Each telescope was sitting in a different part of the world, one more northern than the other, looking out at a different part of the universe and covering most of the night.

I had to get the optics coated in Canada, on Vancouver Island, and then take them on to the US, going from one workshop to the next to make sure everything was ready before they were sent to Chile. As I went through American customs I was stopped, and they wanted to know what was in the large one-metre-diameter boxes. I said they were optics for a large telescope in Chile. 'Well, you look clever so we will let you through,' they said. I didn't think I looked clever – I was quite knackered from all the travelling – but I suppose it seemed like a pretty outlandish story to make up.

They needed someone to oversee bHROS being set up at its final home and, as the systems engineer, I was the only person with enough extensive knowledge. Everything was taken apart and carefully packed in boxes by Pickfords, who were experts at this kind of move. The instrument was 4.5 metres long, 2 metres tall and about 2.5 metres wide. Most of the kit sat on a huge optical bench. It all went to Chile in 27 specially manufactured crates, and my job was to put it back together when it arrived.

CHAPTER 8

The boxes were sent out to Chile on a freight aircraft while I travelled in a passenger one. But I was able to beat the precious cargo, and I watched with relief as each of the 27 boxes was unloaded from the aeroplane and put on a lorry to be driven up the mountain along small rural roads. Years before, to transport the telescope's mighty eight-metre mirror made of solid glass, they had widened some of the tunnels. The route to the telescope was breathtaking as we wound up above the clouds.

I was back and forth to Chile over a period of nine months. It was one of the happiest periods of my life, working on one of the largest telescopes in the world. I had a little bungalow on the adjacent mountain with stunning views of the Andes. Once a week, on a Saturday morning, I would travel down to La Serena, the nearest town to the telescope, to get supplies from the vast supermarket. (This always included a lemon meringue pie and a bottle of Chilean champagne.) I spent every Saturday night in a hotel there by the sea, and then on Sunday morning I would pack my stuff and head back up the mountain ready, for another week of work on the telescope. I was working with a Chilean team and many of them didn't speak much English, though what they did speak was far superior to my very limited Spanish. We managed to communicate through minimal sentences and lots of drawings. They were great guys, and were kind and helpful to the intense Black lady who was obsessed with her instrument. I really enjoyed spending time with them.

MILESTONES

We would work long hours and then I would navigate my way back down the mountain in the pick-up truck I had hired, bleary eyed on too little sleep and trying not to fall off the edge of the track. I encountered a lot of wildlife in Chile, including a tarantula at the side of the road and a small white scorpion which had snuck into one of the packing cases. Once I was chased by a donkey who seemed to take exception to me, galloping after me without giving any sign of stopping. I was also chased by a strange animal, I later found out that it was called a viscacha, but to me it looked a like rabbit on steroids, and I hallucinated it wearing an eye-patch with a knife in its mouth, snarling and saying 'I'm going to get you.' It was at times like this I knew I hadn't slept enough, and I needed to go to bed.

My first job on the telescope was to reconstruct the instrument. It was like starting the ultimate 3D jigsaw puzzle. Firstly, I needed to put the optics into the housings and get everything carefully in place. It sat in a large temperature-controlled box to make sure there were no spurious readings from heat fluctuation. Once it was stable we followed a series of tests, making sure the fibre optics worked with the telescope and could feed the light efficiently into the bHROS. This was a challenge in itself. It was a mind-boggling experience.

I was on the project for around three years in total, before we handed bHROS to the scientific community to use it to make great discoveries about the universe. I spent

CHAPTER 8

time there with the scientist attached to the project so we could utilise the instrument for science, and got my name on some papers, which was a nice ending. They didn't use the technology for as long as I'd hoped, but it served the purpose it was built for. I wish I had gone back one last time because I was so involved in it, I used to see it as my child. It was my baby.

Chile remains one of my favourite places in the world. Occasionally the software guy or some of the mechanical engineers from UCL would come out and we would work and then go off on forays, including into Patagonia. The country is vast and diverse, with the Andes on one side, the Pacific Ocean on the other, Antarctica at the bottom and the Atacama Desert at the top. It is such an amazing place. I found it almost impossible to leave the mountain and the telescope, so much so that Martin had to come and get me and bring me home, otherwise I would probably still be there now. It's hard to describe just how beautiful it is there; the sunsets blow your mind, and the stars make your heart sing.

It also has a wonderful indigenous wine supply, and I would always pick up a couple of bottles for those evenings after work when I would sit, toast the moon and study the stars. I wasn't lonely there at night; I realised I was quite good in my own company. I couldn't believe my luck. What would little Maggie make of it all?! From working on my own mirror of about 150 millimetres in diameter as a teenager to the eight-metre mirrors of Gemini: I wished I could go back in time and tell her what was ahead.

*

When I was at university, one of the largest telescopes in the world was the William Herschel, which clocks in at 4.2-metres, based in La Palma in the Canary Islands. At the time, we couldn't have telescopes much bigger than this because of a phenomenon called atmospheric turbulence, similar to the sort of thing someone might experience in an aeroplane. This turbulence causes the stars and other bodies we see to twinkle. As romantic an idea as twinkling stars are, they play havoc for astronomers who are trying to image objects precisely without them moving around. As starlight travels across space, it's beautifully steady. But the moment it enters Earth's atmosphere, it has to pass through layers of moving air. These layers have different temperatures and densities, which bend and wiggle the light in slightly different directions. So, by the time that light reaches our eyes or telescopes, it looks as if the star is shimmering, flickering or even changing colour.

It's a bit like looking at a coin at the bottom of a swimming pool: the coin is still, but the rippling water makes it appear to wobble and dance. That's what our atmosphere does to starlight!

CHAPTER 8

The turbulent pockets are around four metres in diameter, so if you build a telescope larger than this you are likely to be imaging through multiple disrupted pockets. If we wanted bigger telescopes we needed a way of smoothing out the atmospheric disturbances.

This is where adaptive optics technology comes in, which was first used on a telescope in 1989. The system will image an object in the night sky, let's say a star or sometimes an artificial star generated by a laser on the telescope. By analysing the light from this object the adaptive optics system can work out how that star is twinkling in response to the atmospheric movement. Then we take that response and feed it back into the telescope, but inverted, distorting one of the smaller mirrors in the telescope many hundreds of times a second to take out the twinkle. We can actually detwinkle the stars.

It's basically cosmic noise-cancelling headphones for starlight! Just as headphones sample the background, for example the hum of a plane, and create the exact opposite sound to cancel it, adaptive optics measures the atmospheric wobble and flips it, so the star's light arrives crisp and clear.

With adaptive optics, we can use a much bigger telescope as we are no longer limited by the size of the atmospheric pockets. So, an image that would move all over the place without adaptive optics becomes a nice steady point of light that we can analyse in detail.

Since 1989, many more large telescopes have been constructed, including the eight-metre Gemini telescopes and an amazing group of telescopes called the VLT (which stands for – wait for it – very large telescopes). This system consists of not one but *four* eight-metre telescopes, which is great in itself, but what is even better is that we can combine the light from each of them to make the equivalent of one ginormous instrument.

We combine their light using a technique called interferometry, similar to what I did for my PhD but on a much grander scale. Instead of just taking images and adding up, they add up the *waves of light* themselves. To do this, the telescopes have underground tracks where the light from one telescope can be trapped for a fraction of a second before it's combined with the light from another telescope. This allows the crests and troughs of the waves to line up and interfere with each other, and from that interference we can extract incredibly fine detail – or, as we astronomers call it, we get better resolution.

The effect? It's like having the vision of a telescope as wide as the distance between the separate telescopes – a super-eye on the sky!

Thanks to this trick, the VLT can zoom in on distant stars, watch dust swirling around black holes, and even just about see the atmospheres of exoplanets. It's astronomy's equivalent of a giant cosmic zoom lens, built not from glass, but from teamwork and clever physics.

As pioneering as the VLT is, it is now over 25 years old, so we are looking at building the next generation of

CHAPTER 8

ground-based optical telescopes, pushing the size of the mirror to gather more light and improve the resolution further. There is a new telescope called the ELT which stands for extremely large telescope. (I promise I'm not making this up.) The primary mirror is 39 metres in diameter. Lori and I went out to see it being built a few years ago, and it blew my mind. It's nearly as tall as Big Ben and the size of a football stadium. We hope that it will have its first look at the cosmos, a moment we call first light, sometime in 2029. It will be phenomenal.

As astronomers we are always looking to build a bigger, better telescope, and that is where a telescope named the OWL comes in (it stands for overwhelmingly large telescope). This one is just a concept so it may never get built, but if it was given the go ahead, it would have a primary mirror 100 metres in diameter.

So, if we can do all this glorious astronomy from here on the ground, why are we putting telescopes into space? The answer lies in the electromagnetic spectrum. Stars like our sun shine brightly because light is generated in their heart due to a fusion process. Fusion happens in the super-hot centre of stars where the temperatures and pressures are off any Earthly scale. In this cauldron, the atoms, which are the fundamental building blocks of all matter, get fused together. Simple elements like hydrogen and helium are smashed together to make the atoms of more complex elements. As this process occurs on a grand scale within the star, huge amounts of energy are released in the form of light – visible light that we are familiar with, but also lots of other light such as radio waves and infrared. We've also got the high-energy stuff like gamma rays and x-rays. All these different types of light are classed as the electromagnetic spectrum, and they are one of the few forms of energy that can pass through the vacuum of space, which makes them great for astronomers to use for observations. Other waves, such as sound, need a medium to travel through; this means that the old adage is true – in space, nobody can hear you scream. There is nothing to carry the vibration.

Different parts of the electromagnetic spectrum pass through the vacuum of space, but our atmosphere blocks out some of the spectrum – which has its benefits as it protects us from the harmful rays – but it also means that we cannot study all of the wavelengths generated by stars and other objects from here on Earth's surface. To capture this light, we need to put telescopes above Earth's atmosphere. We need to put these telescopes in space. Enter the Hubble and the James Webb telescopes.

Hubble is a telescope that gathers visible light, and it has sat in low-Earth orbit (around 500 kilometres above sea level) for over 30 years. In that time it has taken many amazing images, including some that have given us an understanding of our insignificant place in the universe.

When Hubble was first launched it was incredibly embarrassing because the mirror was manufactured to the

CHAPTER 8

wrong specification, so when the first images came back, they were horribly blurry. But in 1993, astronauts flew up on a space shuttle to perform a daring repair mission and gave Hubble what I like to call a pair of cosmic spectacles. They installed a clever system of small mirrors called COSTAR. Each of the mirrors was around 20 millimetres in diameter, but together they were able to bend and correct the incoming light so it reached Hubble's instruments in perfect focus.

The fix worked, and the fuzzy smudges turned into crystal-clear galaxies, nebulae and star clusters. Hubble went from an embarrassment to a triumph, transforming our view of the cosmos.

It's a brilliant reminder: science isn't just about getting things right the first time. It's also trial and error; it's about fixing problems with creativity, teamwork and a dash of daring. And thanks to those astronaut 'opticians', Hubble's dazzling vision has continued to inspire the world ever since.

As great as Hubble has been, it still only works in one main waveband around visible light. And some of the questions that the Hubble telescope was throwing up needed an infrared telescope to gain clearer answers. There are some parts of the universe obscured by dust and gas, which visible light can't penetrate but infrared light can. This is the job of the James Webb space telescope (JWST), part telescope, part time machine, searching for the first light from the first stars and galaxies of the universe. JWST is an infrared

MILESTONES

telescope, so it picks up heat energy from stars, galaxies and other bodies. It was designed and built by 3 space agencies, 16 different countries across the world and a total of 10,000 scientists and engineers, and I am very proud to count myself as one of those scientists/engineers. We came together to work on this magnificent beast in order to gain a better understanding of the universe.

The JWST is the biggest space telescope ever built, so getting it into space was no mean feat. It weighs six tons with a mirror 6.5 metres in diameter, and it had to be folded up like origami to fit on board a spacecraft. Once the telescope was launched into space, it moved further and further away, as 300 mechanisms were deployed before it could start working. It was a nerve-wracking time as the instrument unfurled in the darkness, particularly for the 6.5-metre mirror, which was folded in on itself.

While Hubble sits in low-Earth orbit, JWST sits 1.5 million kilometres away from our planet, looking away from us and the sun into deep dark space. If anything goes wrong with it, there is no going back. We can't get out that far to fix it.

The speed of light in a vacuum is 300 million metres per second. It is the fastest known thing in the universe. But although it is fast, it is finite. To understand this, let us consider our local star, the sun. If the sun winked out of existence, we wouldn't know for about eight minutes: that's how long it takes for the light to come from the sun and hit Earth.

CHAPTER 8

Now, the sun is local to us, but we've observed objects in outer space from which the light left 12 billion years ago – it's taken that long for the light to reach us. Some of the infrared light detected by the JWST began its journey at a time soon after the Big Bang, and yet we are only picking it up now. This means that this light is showing us the universe as it was when the light started its journey, some 12 billion years ago. In this respect the JWST is a time machine, picking up light from the early cosmos.

*

Spending months living on a mountain in Chile was exactly what I needed to help heal my broken heart. After productive days, I sat outside every night with a glass of wine, feeling the companionship of the moon and how close I felt to my father, even though I was thousands of miles from home, and he was now in a place I couldn't reach. Whatever I believed about God, if heaven existed, I was certainly closer to it there than I was in Surrey. I did wonder if I would get bored of looking at the stars every night, but that never happened. Every time, I was in awe of the night sky, and I would go to bed just wanting more.

At the end of the project, before I left the telescope, I wrote about my father on a sticker which I put on the underside of the very large optical table, dedicating the project to him, as my inspiration. It was recognition that I would never have got to this epic moment in my life without him.

The other milestone in my life that my father did not get to see was my wedding day, although I like to think he knew because of the way Martin proposed.

In a little village called Compton, near where we live, is a beautiful cemetery with a tiny Arts and Crafts building called Watts Chapel. After my father died, I would often visit – it was a place where I could think about him and feel connected, even though he wasn't buried there. The chapel was decorated as an artistic collaboration by many of the local villagers, overseen by the Victorian designer Mary Fraser-Tytler in memory of her artist husband, George Watts. The walls and ceilings are covered in incredible murals and the landscape is restful, so it is a tranquil oasis and somewhere I have always felt closer to my father.

Martin proposed in the cemetery. That may sound strange, but it was my happy place and really close to my heart. In some way, it felt as if my father was bearing witness to this proposal, and I think he would have given his blessing. I know he was sad that he had not had grandchildren and, while his name continues through my cousins, one of whom is the head of the Yoruba tribe in Nigeria, there was no direct legacy.

By this point, Martin and I had been going out for about twelve years, so we certainly weren't rushing into anything. We planned and paid for the wedding ourselves,

CHAPTER 8

with a little help from his parents and my mum. It was eye-watering how expensive everything was – venue hire, food, drink – and we discounted a lot of options. Then we went to the venue where my university friends Sue and Baz got married, a place called East Hampstead Park. It was truly lovely and, we had assumed, well out of our budget, but it turned out to be council-run, which meant it was more affordable than we realised. Martin and I married there in the summer of 2002. There was a mix-up with dates: our June choice was double-booked, so we moved it to July. We were disappointed at the time but it turned out to be a blessing: on the day we were supposed to get married we were helping a friend move a greenhouse and it poured down with rain. I remember saying to Martin that we could have got married in this, as water dribbled off my hood and down my face. On our new date we had brilliant sunshine.

I struggled to find a wedding dress I liked. I wanted something streamlined, chic and elegant, and there was nothing like that at home that I could afford. But then I came up with a crazy idea. In the past I had purchased some made-to-measure clothes in Hong Kong and Thailand, and they were extremely good value. Could I do the same for a wedding dress? I had been planning to travel virtually around the world to get the optics coated for the bHROS. What if I modified my ticket to stop off in Thailand and have a wedding dress made there to my own unique design?

I checked, and there was no extra cost for the ticket, in fact it would be slightly cheaper. So I scheduled another trip.

The trip was tight on time, and the focus was on getting the optics coated, but luckily I found a wonderful dressmaker and we came up with the design for the dress, which they were able to make in just three days. It had spaghetti straps, a bodice and a chiffon skirt, trimmed with purple silk (my favourite colour). Now I had two different sorts of precious cargo to transport home: custom-made optics boxes, some containing three-quarters of a metre diameter of solid glass, and a wedding dress.

On paper the journey looked doable but, in reality, it was brutal. The journey back was chaotic, involving multiple flights. The time-zone changes meant there was little opportunity to sleep. I was so tired that I left my passport at the check-in desk of one of the flights; fortunately they were able to send it through on the next plane. On another flight I forgot to collect my wedding dress, which had been kindly stowed away for me. Thankfully, someone ran after me with the dress.

At one point, because of the changes in time zone, I had a 48-hour St Patrick's Day. For two days people said, 'Top of the morning to you!' and by the end I was ready to shout at anyone that mentioned it. St Patrick's Day has never been the same since.

It was a ridiculous idea to get my wedding dress made abroad in the middle of a punishing work schedule, but

CHAPTER 8

I love this sort of puzzle, when something seems impossible and then I discover I can make it work. I was thrilled with the dress, and couldn't wait to wear it.

The hen night was befitting a Londoner. I rented a narrowboat near King's Cross so we could potter up to the Camden Palace for a big night out, just like the old days, and moor up overnight. I have always fancied living on a canal boat in London, so it seemed like a great chance to give it a go.

I sent an email out to family and friends, inviting them to join me for dancing and gentle cruising. Spell check changed 'gentle cruising' but because of my dyslexia I hadn't realised it now read 'genital cursing'. I think some of my friends thought this was some sort of African tradition they didn't know about.

There were about ten of us on the boat including my sisters, my mum, friends like Philippa from university and a couple of Martin's female friends too. It was quite a mixed bunch. My two older sisters are both six foot tall, and I have always moved like I am too, although I'm only five foot seven. They are also quite bombastic. I used to think I was shy and retiring, but this was in comparison to them and to my mum, who is quite the character. Turns out I am quite bombastic too. My sisters tried to get me very drunk, but Philippa, (who had saved me from the dick-machine spelling disaster in my PhD thesis), stepped in yet again and had my back. It was such a fun night and, as we got back on the narrowboat, we saw someone peeing into the canal and my sisters started heckling him. So, in a way, genital cursing did happen.

The night before the wedding, Martin and I stayed in separate rooms in the hotel where we were getting married. As I am always late for everything, it seemed sensible to get ready where we were getting married and avoid tardiness. The hotel staff said they would give me a call when it was time, which took the stress off and meant I wouldn't have to constantly check my watch – except unbeknownst to them and me, the phone wasn't working. By the time we all realised what had happened, I was late for my wedding, even though I was on the premises and ready.

We had the best time. My bridesmaids, in purple silk dresses, were Martin's niece, Emily, and Sue and Baz's little girl, Maggie; she was named after me, which was such an honour. Many of our friends had children by now and we wanted them to feel just as much a part of the day as the adults did, so we arranged for them to have colouring packs at the table and had a bouncy castle in the grounds. In the evening, the adults jumped around on it, including me in my wedding dress.

I missed my father's presence at the wedding, but then I missed him every day anyway, and I gave a short speech talking about him, saying although he wasn't there, he was in our hearts. My auntie was very much there though and, as well as the weird conversations she was having with people,

CHAPTER 8

she was demanding that Martin and I have kids straight away because of my advancing years, almost clearing crockery and glasses off one of the tables as if we were going to have sex on the tablecloth. 'Get to it!' she instructed.

It's a risk getting family together because there always seems to be a feud of some sort, but it was important and made the day feel like the biggest celebration of togetherness. We commissioned an artist to draw a picture of the venue and asked our guests to write messages and sign their names on it, which was a lovely memento of the day.

I was so happy wearing a dress I had designed myself, and I matched it with some stupidly high-heeled black lace-up ankle boots. I don't know how I managed to wear them all day, but I did, and then for the evening I decided to change into something I could dance in as we had booked a live band. I took the chiffon skirt off to reveal a white mini tennis skirt and swapped the teetering heels for a pair of white Doc Martin's I found in Camden Market. We partied into the early hours and then Martin and I wandered up to my hotel room. The next morning my mum was banging on the door asking, 'What are you two up to?!' (I sometimes wonder about the women in my family. OK, often: I often wonder about the women in my family, sometimes including myself.)

The day after the wedding we went to Rome for our honeymoon. To keep the cost down we had booked a small hotel quite close to the train station. We came back one night after a delicious evening full of pasta and I was wearing a bodice-style top and midi skirt. The guy on reception said to Martin, 'Whoa there, you can't take her up to your room.' We had to explain that we were in fact on our honeymoon, and I was not a lady of the night.

Hal had also treated us to a weekend away in Barcelona as a wedding present, which is another of my favourite places on earth, and we had a brilliant time there too. Both honeymoon trips were full of good food and wonderful architecture, two things I continue to look for in every adventure.

*

As my personal life took a new step forward, so did my work. I had completed my mission in Chile and returned to UCL. When I had been taken on, I had no experience of managing space instrumentation, but I had now proved my capabilities and was absorbed back into the MSSL. This meant I had made the transition from ground-based astronomy to space science. I had officially become a space scientist and achieved another of my crazy dreams!

I was part of a team putting in a bid for instrumentation on a new high-tech telescope. It was called the Next Generation Space Telescope, which really sounded like something from *Star Trek*. Soon afterwards, it was renamed the

CHAPTER 8

James Webb space telescope, in memory of the influential second administrator of NASA. I was working on a concept for an infrared spectrometer which we called NIRSpec, with a large industrial company. As UCL is a university and this was a multi-million euro project, it wasn't something we could do on our own, but if we collaborated with industry, we had more chance of success.

The company was based all over Europe and I was travelling to Cannes every week, spending several days there at a time. I really enjoyed being part of the team, working with a large company and putting a bid together. We had to wait for the ESA to decide who they were going to go for, which turned out not to be us. Little did I know then, but years later I would get a chance to work on one of the JWST instruments.

I was looking for the next challenge, and applied to be a space manager at a private company, Sira (which stood for Scientific Instrument Research Association). This felt like a good move. I knew I wasn't destined for a job that was dependent on my written work. As I mentioned earlier, academic science jobs often involve a lot of writing, which was one of the reasons I went into industry, certain I would thrive better there.

The interview for the job went well – so well that one of the people (let's call him John) on the panel fired many questions at me. I assumed this was because John was interested in my previous experience, and I loved nothing better than talking about my projects, so I could have sat there for hours. I got the job, but it was only much later that I discovered why he had interrogated me so thoroughly. Around two years after I started working there, I was given my personal records, in which he had written that I wasn't suitable for the role, with no real reason to support this. He had been digging and digging in the hope that I would fall over.

The MD of the company had been uncomfortable about John's cross examination too, and apologised some time later. He noted that John had asked me a total of 42 questions in the interview, and was surprised by his zeal. It didn't stop me being furious as I read my notes. John was damning, concluding that I wasn't management material, and I wouldn't go very far. He had been actively trying to stop my progression in the company, and I couldn't understand why he had taken against me so much. There were only three possible reasons:

1. He really didn't rate me.
2. I was a woman.
3. I was Black.

I can deal with flagrant racism, but when it is behind closed doors, infiltrating the network without me being aware of it, then it's much harder to respond to. This experience was a real eye-opener for me at this point in my career.

CHAPTER 8

Experiencing blatant hatred because of the colour of my skin has been thankfully rare. It can be more shocking for those around me. One evening, Martin and I went out for a drink with our friends Paul and Paula. We were sitting by the river when a group of strangers began hurling racial abuse at me. Paul and Paula were gobsmacked by the behaviour and had never witnessed this before. Luckily, it's not a world I see often either and, when I do, I see it as the other person's problem not mine.

What I find harder is the racism that you can't see. It is difficult to tackle something when it is not in your face and there is nothing tangible to hold on to, or for others around you to notice. The South African comedian Trevor Noah calls it 'charming racism', the sort that hides in plain sight, insidious and lethal.

It was impossible for anyone to misconstrue the incident down by the river, but what about walking into a building in a suit and someone assuming I am the cleaner, or being in a meeting room where someone asks me to get them a coffee? Being at the BBC for an event and a female senior executive asking me for a glass of water and then, on realising I was not waiting staff but part of the delegation, avoiding me for the rest of the evening? If I didn't get a job, is it because I am not right for the position? Or is it because I am female? Or Black? Do I just not fit in?

These are the sort of moments that can play with your mind. Some people describe this as death by a thousand cuts. I am also very aware that I experience a lot less prejudice than many, and I am in a privileged media bubble surrounded by similar-thinking people. It's easy to seek out people who think the same way you do, in an echo chamber of agreement that doesn't represent the wider world.

I watched a programme about racism in the Deep South of America with Martin years ago. The presenter was travelling around, speaking to people, remarking on how courteous and accommodating everyone was and how much they made him feel at home. And then he said something that hit a nerve: 'At the end of the day, you are never really sure who is wearing the white pointy hats' – a reference to the Ku Klux Klan. From his perspective, as a white man, everything was fine, but had he been Black, the situation may have been different, and the hospitality may not have been there. I realised this is how I had felt, sometimes consciously, for most of my life. It resonated and reminded me of the fear I carried from childhood, knowing some people would not be able to accept who I am.

When I went to meet Martin at a conference in Texas, I wasn't sure, as a Black person, if I could just turn up at the hotel and say I was meeting my husband. He couldn't understand my reticence, because from his perspective he had been welcomed and everyone loved his British accent. He assumed it would be the same for me. The point was, I carried fear into every situation and was ready for something to go wrong. It didn't, but it could have done.

CHAPTER 8

Martin and I used to play this made-up game called 'Be Careful, You'll Crash!' because in some parts of the world, I may be walking along and someone would drive past, clock me and be visibly shocked to see a Black person. In South America, someone came up to me and actually scrubbed my face because they thought I was wearing make-up.

I talked to Martin about it. Sometimes when something happened or I had an uncomfortable feeling about a situation, he'd say it wasn't that bad or want me to see the funny side, which I often do, but what I needed from him was to empathise more often, to agree with me and and say, 'Yep, that was horrendous'. He was trying to make me feel better, but instead I felt completely isolated. I wanted us to look at the bad thing together, acknowledge it and call it what it is, thus diminishing its power to hurt me.

I did an event for the French Institute in South Kensington quite a few years ago, and I was chatting to the organisers before we started and asked why they chose me. They said they needed someone of a different ethnicity, plus the panel was all men so they thought they should ask a woman. It was clear that I wasn't invited just because of my skills, but to tick a couple of boxes. In some ways it was quite refreshing that they were so open about the situation. I admired their honesty. But it was awkward, and a bit insulting. They didn't dress up the invitation as something else. That said, I wasn't just going to behave like a gun for hire.

I chaired their panel discussion with the heads of the

UK and French space agencies and the astronaut Tim Peake. It was around the time of Brexit, and just before the end of the session, as they were wrapping up, I said I had one last question to ask which wasn't on the programme. I asked, 'How is Brexit going to influence the space industry in the future?'

That put the cat among the pigeons! A debate ensued which I think was useful and informative for everyone in the room. Furthermore, I showed that I was not just a puppet who read out questions, but I was there as an audience representative to raise the crucial points. I will always step out of the box some people put me in.

*

I was invited onto *Newsnight* to talk about a scientific discovery associated with the origins of the universe, along with the fantastic Sri Lankan astronomer, Hiranya Peiris, another female scientist. It was a complex story and the scientist who had made the discovery was on the line talking about it. Jeremy Paxman suddenly turned to me and said, 'Why am I asking all the questions? Why don't you ask a question, Maggie.' This completely threw me. I was not prepared. I was very happy to comment on the findings, but to ask the scientist about their work, particularly as it was not in my field, put me on the back foot.

CHAPTER 8

I remember feeling awkward, and there was a little nervous chuckling, but we got through the interview, and I thought no more about it until a piece ran in the *Daily Mail*. It was written under a pseudonym in a weekly column, and made a comment about me giggling and wondered why two women had been asked to talk about a report about white, male American scientists, the implication being that the BBC had drafted in ethnic minority women to talk about the work of men.

As is my way, I tried to make light of it with a comment about the BBC flipping through their Rolodex in search of diverse panel guests, but I made the point that I believed I was asked to appear on the programme for my ability to translate complex ideas into something accessible, rather than for my gender or the colour of my skin. Professor Hiranya Peiris followed up with the damning disappointment that within this debacle the journalist had also 'erased the contributions of all of the non-white, non-male and non-American scientists involved in the discovery'.

This sort of situation was familiar to me, and I decided to rise above the snide article and not let it trouble me. But for the first time in my professional life, someone stood up for me.

UCL read the piece, and they were not having any of it. They were the first university in the world to admit women and they had associations with both Hiranya and me, so they decided to come out fighting on our behalf. They instigated a campaign and wrote an open letter to the editor of the *Daily Mail* about the 'profoundly insulting' piece. They said, 'It is deeply disappointing that you thought it acceptable to print an article drawing attention to the gender and race of scientific experts, suggesting that non-white, non-male scientists are somehow incapable of speaking on the basis of their qualifications and expertise.' The BBC commented too: 'We ask people onto *Newsnight* because we think they know what they are talking about and have something interesting to say.'

The *Daily Mail* issued an apology to us. OK, so it was a few sentences printed in a tiny corner, buried many pages into the paper, but it was there in black and white. They also sent me a more fulsome letter of apology, which I framed and is hanging on the wall of my downstairs loo.

*

As much as I have battled stereotypes my entire life, I realised quite recently that I had unwittingly made similar assumptions about someone I had never met.

In 2019, I was invited to the premiere of the biopic *Armstrong*, which was made to coincide with the anniversary of the moon landing. I sat on a Q&A panel with Neil Armstrong's son, and Neil's granddaughter sang one of the songs from the film's soundtrack. Being in this environment

CHAPTER 8

gave me a better insight into Neil Armstrong. I had him pegged as an adrenaline-fuelled space jockey with a buzz cut, standing on the pinnacle of white, male, testosterone-powered dominance, but I had got him all wrong. My initial image of him had come from what I imagined the 'right stuff' was all about. Images portraying the alpha males of the 1950s vying for positions at the top of the heap. I assumed that, because he had got arguably the most coveted position of all time, he must have exhibited the characteristics listed above. This belief was probably fuelled by the sci-fi I read growing up, which, as I mentioned before, could be misogynistic. In watching the film and discussing his life with those who knew him best, his family, I realised he was a quiet, mild-mannered engineer, an introvert who was thrust into the global spotlight.

Armstrong had proved his mettle in an earlier mission, when his logical thinking in an incredibly stressful situation, both physically and mentally, had averted disaster, saving his life and that of his fellow astronaut. His selection for the moon landing was based on his ability to stay calm under pressure, focus on his job and not chase the glory of the incredible event. It also gave me a deeper respect for the space mission, because NASA was not just throwing a group of astronauts up in space to beat the Russians; they had carefully found the right people for the job.

Just before Armstrong joined the space programme, he lost his two-year-old daughter to cancer. It is rumoured that he left her bracelet on the surface of the moon as a memorial to her. I like to believe he did; it seems like the sort of thing he would do. After the international attention around the mission, Armstrong quietly faded into the background. Unwilling to become a celebrity, he was considered a recluse, but he had been as brave on Earth as he had in space.

My entire perspective changed. Armstrong reminded me that judging people based on stereotypes is never a good idea. As a Black female scientist working in a white male-dominated domain, I have felt the sting of such assumptions myself and thought I would be less likely to fall into the same trap, but I now know that I must remain vigilant in my own attitudes.

*

As a space scientist, I haven't spent my career just focused on what lies beyond our planet. Much of the work I have done has been designing and building Earth-observation satellites, designed to look at our planet – things like the Disaster Monitoring Constellation, a collaboration between countries across the world, each with a satellite that sits in low-Earth orbit. When a disaster happens, the satellites take images as they pass over and send them down to NGOs and government bodies, giving a global view from space. Data like this make a difference to the world. Sometimes people

CHAPTER 8

ask if we be should spending all this money on space, but most of the money we spend is designed to help us on our own planet.

At Sira, I was tasked with a number of different projects including one called Aeolus, named after the keeper of the winds from Greek mythology, because we were measuring wind speed in Earth's atmosphere. The wind connects weather cells across our planet, which gives us valuable foresight and a better understanding of climate change.

Being a project manager is like playing a game of chess, especially when resources are limited, and to build any instrumentation needs a number of different disciplines. It takes a village. Or at least a big team including designers, mechanical engineers and software experts who need to collaborate. Often, they are called to work on other projects, which raises the jeopardy and heightens the negotiation for resources.

I enticed everyone to my project-management meetings with a big bag of doughnuts. Yes, sometimes it really can be that simple. It also helped break the ice, as I sensed some people wondering whether I could handle the job. I had not yet read John's damning report, but I did have a conversation with another colleague, a friend of mine, who had said how brave Sira were for hiring a Black woman. 'But if I am good at my job, if I am a competent scientist, why is it brave?' I asked. So, yes, I was an unexpected choice as project manager, especially in the space industry, and I knew that I had some barriers to break down.

MILESTONES

A successful project manager is, I believe, someone who will do anything to help the project happen. They will take on tasks that may not be directly related to their job, which could be anything from working long hours to emptying the bins – if it smooths the road of the project, that's the job. That's how I saw it. It showed others how important the project was to me, and that I would do everything in my power to help the team and create the right environment for them to achieve and thrive. If I was asking people to step up, I had to lead by example.

In many ways, project management was the perfect position for me, but it did threaten – and often topple – my work–life balance. I have a workaholic tendency to throw my entire self into anything I am working on, and it becomes all consuming. This is less of a reflection on the task and more on my nature, which now makes more sense with my ADHD diagnosis.

My job was based in Chislehurst, near Sevenoaks, so I had to commute from Surrey every day. Martin and I had just bought our first house in Guildford. This had been more my idea than his, because I was very keen to get on the property ladder. We found a 1930s semi-detached house, with a shared drive. Martin was settled in his job for an engineering consultancy. After his PhD, he had stayed on to do postdoctoral work. With a numerate degree and PhD, there were quite a few opportunities in the city, but he didn't want to take his career in that direction, so

CHAPTER 8

he took the consultancy route and stayed there for over 20 years.

This was anathema to me. I liked moving around and looking for new jobs; it seemed like a natural thing to do, allowing me to evolve. This was unlikely if I was stuck somewhere, because I wasn't pushy and tended not to apply for promotions. However, if I was transferred to another group or job, I found it easier to climb to the next rung of the ladder. It meant I had to move jobs to be able to progress in my work. This feels reminiscent of me changing schools so many times and reinventing myself with every move. It has been in me from an early age, and I can recognise the feeling of itchy feet, the need to redefine who I am and what my goals are.

It was 2005 and I was settling in at Sira, but another, more personal project was tugging at my sleeve: my passion for science communication. I entered a competition, FameLab, run by the Cheltenham Science Festival, which promotes the best new voices in science communication. I got through to the UK semi-finals before I was knocked out. This was really disappointing, but it didn't dampen my enthusiasm for spreading the word, I just had to find another way. I wanted to get out there to make the scientific community more diverse, and science communication seemed a great way of doing this.

I wrote to schools, introducing myself: 'Hi, my name is Dr Maggie, I'm a space scientist and I would love to come to your school and give a talk.' You know, the sort of thing that elicits mostly silence, but the occasional positive response. I was new to the game and untested on the circuit, so I knew it would be a gradual growth. I continued to work full time on the day job while this was on the back burner, and made the occasional foray to a local school to speak to students.

As luck would have it, a government organisation then called the Particle Physics and Astronomy Research Council (PPARC) was offering grants as part of a media outreach scheme. PPARC was the organisation that had funded the construction of bHROS, so I was not a complete outsider. I applied for the funding, and was offered £6,000 to support my plan around science communication and buy the demo equipment I needed. The money had to be paid into an auditable company, so I set up Science Innovation Ltd in 2004. With a business and a little money behind me, I could approach schools on a more official footing and my reputation grew, although not always in the right way. I did manage to slime a school once because I was showing the students how to make slime out of household substances, and they took it out of the classroom and around the school. Everywhere you looked, there was more slime. I wasn't invited back.

I ran my two commitments together, establishing my passion project alongside my intense day job, which occasionally caused friction if I had booked a day off to visit a school and it clashed with a vital meeting at work. I always took annual leave for Science Innovation commitments, so

CHAPTER 8

it was in my own time, but when a project hits a critical phase it's hard to walk away for a day. Sira knew about and supported my sideline, and the big bosses said how marvellous it was that I was inspiring the next generation, but my immediate bosses were more aware of the potential timetable clashes and often not so keen.

While it created a little tension sometimes, I wasn't prepared to give it up because I loved speaking to kids. I told them about little Maggie wanting to become a space scientist and how unlikely this was to happen, and yet here we were. I would talk about looking for opportunities and focusing on the end goal as a way to overcome hurdles and hoped, with all my might, that my words would stick to these students and they would carry them through their studies into the future they wanted. Every time I give one of these talks, I think about little Maggie and how, if she can do it, anyone can.

*

Life was pottering along, between Sira and Science Innovation, and after one particular school visit I was summoned back to the office. When I returned, people were being split into two separate rooms, with about 80 per cent of the workforce in one and the rest in another. Something was up. I joined the largest group but was told I was in the wrong room and moved to the other group, where the bosses announced the company was going into administration. The staff in the 80 per cent room had been made redundant and had to clear their desks immediately. The remaining 20 per cent of us would continue working because there was an option for that part of the company to be bought out. It was mainly the space division that had been given a temporary stay of execution.

Sira was based in a stately home, similar to MSSL, which now felt very empty with only a fifth of its previous occupants. The maintenance team had been let go, so as we continued to work in the place it fell apart around us. The garden became overgrown, and we would have to hack our way through to the main house to get supplies from the eerily empty laboratory buildings. It was like the start of a weird horror film and a strange environment to turn up to every day. What's more, although I felt incredibly lucky to still have a job, the clock was ticking. If Sira couldn't sell the space division of the company, it was curtains for the lot of us.

As a child, I used to watch the news even though I hated it. I have always found the news hard work. Most of the stories were quite depressing and I empathised with what I saw, but I would also feel frustrated by my inability to change the general demise of the world. There was a global recession for a couple of years, from 1973, and they would show diagrams to indicate how dire the job market was. I remember a map of the UK, and the thousands of jobs lost would

CHAPTER 8

be highlighted in red circles scattered across the country and then they would show how many jobs had been created – there were so few of them. I thought, *I never want to be in a red circle*. I decided I had to find a way of working for myself and then, if it went wrong, that was because of something I had done. I would be in control of whether it succeeded or failed.

This has had a bearing throughout my career, and particularly resonated while I was working at Sira and establishing Science Innovation. I wanted to avoid putting all my eggs in one basket. A portfolio career was the answer.

Surrey Satellites, based in Guildford, bought Sira several months later, and my colleagues thought I had insider knowledge because we had bought a house there. Of course I had no idea, and nor was I planning to stick around to benefit from the quick commute. For the first time in my life I was headhunted, by Astrium, who were once again putting in a bid to work on the James Webb space telescope and wanted me to set up my own optical group there.

I joined as the project manager, looking at a five-year plan to build instrumentation for the telescope. I was working with a team to put this bid together, but the problem at this stage was to create a workforce without any money coming in, so we could come up with a rough design of the instrument and work out how to build it. It was a major endeavour, and this time a bag of doughnuts wasn't enough to galvanise the troops. Added to this, my boss was signed off

MILESTONES

on medical leave soon after I started, and so I had to manage the multi-million-pound bid on my own. While I had experience from my previous job, being in overall charge was absolutely terrifying. And our direct competitors were my old team from Surrey Satellites – space science is a small world, where you bump into the same people regularly.

To make matters even more complicated, I had taken on a part-time role in my free time. PPARC, the group that had funded bHROS and enabled me to get my foothold in science communication with Science Innovation Ltd., made an announcement. They were creating media fellowships. These would involve working with media outlets, newspapers, radio and TV to promote science to a wide range of audiences, with a focus on encouraging young people to consider careers in STEM. This encapsulated so much of what I believed in and had been working on, so I applied for it. The fund was mainly aimed at academics, so, because I was in industry, I had to be connected to a university to qualify for the grant. I approached UCL and the wonderful Professor Steve Miller, who headed up the science communication centre there. As I had already worked at UCL before leaving for industry it was a good fit, and he was willing to take on the grant in his department, if it was awarded.

My application was successful, and Steve and I would catch up over coffee every now and again to discuss what I was up to. He would kindly make suggestions which I would take on board at the time but often failed to implement.

CHAPTER 8

He affectionately called me 'Mag the Unmanageable' because I always did things my own way. I still do. I think that's why I don't like sitting on committees. I go at things full tilt and have a vision of what I want to achieve, then I just want to get on with it rather than having lengthy discussions. I think that it might be a symptom of my ADHD. When you run at things with minimum planning you know that some things will go awry, but I'm used to thinking on my feet. I'm used to change. I like to update and adjust if needed; I like to be agile and adapt, but I know that this is not for everyone. So the lone wolf in me prowls around impatiently.

One of the skills I needed to possess in my communication job was the ability to stand in front of an audience and captivate them with stories of science. I had a taste of it when I did my transfer talk for my PhD, the examination to determine whether my study so far could lead to a PhD not an MSc, and the feedback I got was how entertaining I had been while still being factually correct. Years before, I entered a public speaking competition while working at the MOD. I won and was excited about my prize, which was billed as a nice bottle of Margaux wine, except I ended up getting two rubbish bottles instead. What I did gain was the sure knowledge that I could engage an audience with my public speaking and, most importantly, I really loved it.

I ignored the standard scientist approach of speaking in the third person and remaining impartial, which works on paper and is necessary to show impartiality, but is less effective when you're in a room full of people, some of whom may be stifling a yawn. I wanted to light up the place and make it a fun experience. I feel the same way now. Maybe it's the undiscovered actress in me; I want to entertain people, and I always have. The class clown lives on. Once I knew I could incorporate this into my career, there was not stopping me.

I found myself wearing several hats, which was the portfolio career I had dreamed of, but the reality took some energy. I still have anxiety dreams about not managing my time well enough during this period. We put in the bid for the JWST instrumentation and we were unsuccessful, which was a big disappointment after such a gargantuan effort. Also, I was the chief coordinator of all the paperwork and writing: not the ideal role for a dyslexic. As much as I had wanted it to happen, I knew it would have taken me away from my fledgling side-business, which I was passionate about. I was given a background task to fill my Astrium time, working on figuring out how to measure how effectively plants were photosynthesising.

When plants photosynthesise they absorb sunlight and the greenhouse gas carbon dioxide. Plant health gives us a real indication of how our planet is doing. And these measurements could give us an indication of how well we could feed the planet. But the real challenge was doing these measurements from space. It was a very worthwhile endeavour,

CHAPTER 8

and it was great to play with ideas because the project was at such an early phase.

In the cosmos of space science, we are used to hiatus periods between major projects, so I didn't panic. Also, with the work now at UCL and jobs coming in through Science Innovation Ltd, there was more than enough to do. And there was also another reason for my calm, meditative outlook: I was pregnant.

Chapter 9
Starchild

I decided that if I hadn't fallen pregnant by my 40th birthday, I would give up. Martin and I had been trying for a couple of years, and nothing was happening. Setting a deadline was important to me because I hyperfixated on things, and I didn't want to get obsessed with trying to have a baby. It's not something I had any control over, and I didn't want to drive Martin and myself mad trying to make it happen. I wasn't even sure if I could get pregnant. My menstrual cycle had always been inconsistent, and I have endometriosis – I was one of the first women in the country to be given the progesterone implant, which was a gamechanger for the debilitating period pain.

A week before my 40th birthday, I discovered I was pregnant. I could not believe it; I was gobsmacked and thrilled. Then 12 weeks later I miscarried. It was devastating; I felt like this had been my only opportunity, and it had gone. The emotions welled up inside me and I would sit in the car – a private place where I could feel the grief – listening to sad music and crying my eyes out. There wasn't anyone

CHAPTER 9

I could really talk to about how I was feeling, although I mentioned it to my mum who brushed it off with an 'Oh yes, it happened to me many times' then, unhelpfully, 'Have you gone for the scrape?' Not what I wanted to hear. I did speak to Martin, but it just wasn't the same for him. He faced his grief in his own way.

I've always thought kids were amazing. They are magical and resilient, carrying so much potential. If I am walking down the street and a child looks at me and grins, I will always wave back at them. Why wouldn't you? How could you not encourage a child?! There is such an innocent joy. I wanted to immerse myself in that world with my own baby. To me, having a child is the ultimate adventure; you are steering someone's life and you want them to grow up as a happy and well-rounded individual.

The doctor asked why we had left it so late to try for children and I added that to the list of unhelpful comments. We couldn't go back in time and do something different. One day, feeling bereft, I went to Guildford Cathedral, which we can see from our house. I was wandering around and came across a garden of remembrance, established to remember children who were never born. It was a hugely poignant moment for me, and a source of some comfort.

In a way the timing was serendipitous because, had I not got pregnant just before my 40th birthday, I would have given up. So, even though I had lost the baby, I now knew I *could* get pregnant, and it gave me a glimmer of hope to continue trying. It felt like a sign. I turned the pain of loss into a determination to keep going and began charting my menstrual cycle. (I am glad I didn't know what I was told recently, that past the age of 40 you have just a 3 per cent chance of getting pregnant naturally. It would have felt like the cards were stacked against me, although I have always played against the odds.)

A year later, I was invited to speak at the Latitude music festival, and Martin and I had a fantastic time. I felt so carefree, dancing in tents and lounging on the grass all weekend. A week later I discovered I was pregnant again. It felt like a miracle.

This time around I was incredibly nervous, holding my breath for the first twelve weeks before I was safely past the point when I had miscarried previously. As an older mother, I had the Down's syndrome screening test, where measurements are taken of the baby inside the womb. There it was, bouncing around on the screen, full of life and refusing to stay still so the nurse couldn't get any details. She sent me home to relax and return later in the day, when exactly the same thing happened again. My baby was full of beans! I had a sneaking suspicion this would set the tone for its arrival.

A couple of months later, Martin and I went over to the States and toured around a bit – our last big holiday as just the two of us. In Arizona, we came across a stall selling Indian fetishes – small creatures carved in stone by the indigenous people of the region. I believe that each carving

CHAPTER 9

represents a spirit animal. We were trying to find one that would represent our feisty baby and we chose an otter because they are clever, fun and mischievous. Little did we know how accurate that would be.

I was a happy expectant mother, even though I was weighed down with fear. I tried to push it as far back in my mind as I could and focused on doing as good a job as I could at being pregnant. I paid attention to all the things I could and couldn't do, including foods to avoid, and then one day Martin heard me whoop from the living room. 'Hey, what's happened?' he asked. 'I have just discovered I can eat peanut butter,' I said happily. It's one of my favourite foods, but I'd heard that eating it in pregnancy may cause the unborn child to develop a peanut allergy. However, the evidence was in, and it indicated that this wasn't the case; nobody was happier to hear this than me. Which was lucky because I had been suffering from a severe loss of appetite in pregnancy, and I could go a whole day without eating. The hospital were monitoring me and I was taking lots of supplements to try and make up for it, but by the evening I would have to force myself to eat and only be able to stomach junk food, like pizza.

When I was around five months pregnant, the United Nations got in touch and asked if I would join a delegation to Syria to speak with children there. I had already been to Israel and Japan with the British Council previously, which were mind-blowing experiences, so it was an easy 'yes'.

STARCHILD

This was just before the trouble in Syria started and four of us went out, travelling around the country and giving talks in various locations including schools where the children had been given information on each of the speakers. They were told I was an astronaut, so they wanted to know what it was like to go into space. I had to disappoint them on that score, but I hope I made up for it with other stories I told.

I was visibly pregnant and feeling a little vulnerable. I wanted to make sure that the baby was safe, so I had taken a small portable ultrasound instrument with me which meant I could listen to the baby's heartbeat. I felt such a responsibility to my unborn child, but also to my own belief that I could carry on as normal. One night we went out for a fantastic Arabic meal and ended up dancing, with people smoking around me, and I felt guilty for putting the baby in that situation. Back at the hotel room I whipped out the ultrasound and for a moment I couldn't hear the heartbeat – then there it was. The fear was real, but I tried not to let it throw me.

We ended our visit in Damascus and went out for dinner on our last night. We bumped into a local in the restaurant who was horrified that we hadn't had a chance to see his city properly so, after we had eaten, he took us for a tour in his minivan. He offered us such generous Arabic hospitality, and we made him take some money from us to at least cover the petrol and some of his time. It was a glorious city

CHAPTER 9

full of history and beauty, and seeing it at night was brilliant for me, the confirmed insomniac.

The next morning, I had to wake up very early, before dawn, to catch our flight home. I stood at the window, listening to the call to prayer, which seemed to hop from mosque to mosque, just as the sun began to rise across the city. This would be my last big foreign trip before giving birth, and it had been magical and thought-provoking in equal measure. I met the most incredible children and talked to them about having big dreams for their exciting futures, little knowing that six months later their country would be in the grip of civil conflict. When the news broke, I remembered those children, their education now snatched away, their futures crumbling around them and wished I had spoken of more fundamental needs rather than aspirational things. Meeting those children is just one of the many reasons that I am very proud to be an ambassador for the International Rescue Committee, a fantastic organisation supporting displaced children in places of war across the world – among many other things that they do brilliantly.

*

I spent a lot of time singing to my baby as it did somersaults inside me and grew bigger, pushing an elbow or a foot, stretching the skin beyond my belly. I guess I should have played it Bach, but instead I sang the songs I had learned in my childhood about 'stories of old' and knights who were 'gentle and brave'. This was one of the parts of pregnancy that I adored, keeping my baby safe and protected inside me. My body wasn't quite as buoyant as I had hoped, what with high blood pressure, lack of appetite and extreme tiredness, but I was being closely and regularly monitored by the hospital.

Juggling my three jobs in the last few weeks of pregnancy was exhausting. I was working as a space scientist at Astrium, managing large projects. I was doing some of my science communication work through UCL and I was also running my own company, Science Innovation Ltd. In addition, I was giving talks to large audiences as part of an organisation called GCSE Science Live. This organisation arranges education events, hiring big venues with a seating capacity of a few thousand and inviting local school kids to come and listen to talks from different scientists. I tell kids that studying physics means learning about everything in the universe, from the tiniest subatomic particle to the outer reaches of the cosmos. Gaining knowledge of the universe is so important, whether you go on to do a degree in the subject or not. I think we need that sort of logical thinking in as many places as possible. I would love to see more scientific politicians, for example.

When science is reported, it's usually positive news, maybe a cure for something, great leaps forward in understanding or finding a vaccine. What you don't see is the many years of

CHAPTER 9

study and research it might have taken to get to this point. We don't talk about the negatives, or all the things that go wrong and don't work, but it's important to get that across.

The majority of the things we try as scientists don't work, but that is a positive in itself because we learn from those missteps. I think it is a metaphor for life, for understanding that things go wrong and it's OK to fail. That is still progress. I like to tell my audiences that success is not about not failing, it's about picking yourself up after failure; how you do that defines who you are. I make that point in the talks I give, particularly to school children; they need to know this for the future. I'm still involved in the initiative, and I love the way it reaches so many students in one place and, as well as the learning aspect, it's such a fun day – for the kids, teachers and us.

I was heavily pregnant, but I was not going to miss my events for GCSE Science Live and I continued at Astrium, even though I was on my feet a lot and the car park suddenly seemed an awfully long walk from the office.

A couple of weeks before my due date, I went to hospital for a routine check-up. My appetite was still low, my blood pressure high and I had developed a rash which I assumed was eczema, so I didn't think anything of it until I saw the doctor and mentioned itchy palms too. This rang alarm bells, and they diagnosed me with suspected cholestasis, a liver disorder associated with pregnancy that can put your unborn baby in mortal danger. The hospital was concerned that this hadn't been picked up earlier, and booked me in for an induction the following day. Feeling a combination of fear (please God let nothing happen to my baby) and excitement (I am going to meet my baby!), I rushed home to tell Martin and pack my bag.

*

Going into space has always been a fundamental dream of mine: so too was having a baby. They both echo the thrill, the worry of something going wrong, the elements of danger as the journey begins, the hope that everything you have learned will stand you in good stead and then, if you are lucky, the realisation of the dream. Like an astronaut sitting on a launchpad, I thought, *I have got this far. I am not going to entertain failure. At the end of this adventure I am bringing everyone safely home.*

When I was in hospital I started vomiting. I don't know if it was connected to my liver dysfunction, but it hadn't happened before. It meant I couldn't take any medication orally because the nurses had no idea how much I may have ingested before I was sick again. It made a tricky situation more dangerous, and over the course of 24 hours I got weaker, so they hooked me up to a saline drip. How on earth was I going to have enough strength to go through labour and push a baby out? Then my waters broke.

CHAPTER 9

As worried as I was about my baby, it was as if I was hovering over my body watching the medical drama unfold, fascinated by everything. I had loved studying biology at school, and I am really interested in how the human body works, so this was right up my street. Back in my physical body, any energy I had left had disappeared. The medical team called for an emergency Caesarean, which was fine by me because I wasn't sure I would have had the stamina for a natural birth. *Just get my baby out safely*, I thought.

Martin got dressed in scrubs, I had an epidural and then was wheeled into the operating theatre, where they put a screen up between my head and my stomach so I couldn't see what was happening. I was fixated on hearing the baby cry, which would mean it was healthy, and everything in me was alert to this. Suddenly, she was there in the room, our daughter, our beautiful Lori, and she gave a lusty scream. Thank God. I wanted to feel reassured by this, but I was terribly lightheaded; I thought I might pass out.

Lori was put in Martin's arms and then he gave her to me. She was so beautiful. *Look what we have made together*, I thought, *this exquisite, perfect, small human*. I wanted to stare at her for ever, but instead I said to Martin, 'Can you take her? I think I might drop her because I really don't feel well.' I was suddenly aware of a lot of medical staff being very busy around me and there was a bit of a kerfuffle.

I had lost a litre and a half of blood. They had nicked a couple of internal things they shouldn't have, and I suffered for it. This was a key moment for me to bond with my baby, but I was dipping in and out of consciousness. I kept thinking I had to hold on to her tightly so she knew she was safe, but I couldn't trust myself to do that. Eventually they sewed me up before Lori and I were taken to intensive care and Martin was sent home.

It was such an alien place to be. I was incredibly weak and faint, but I had Lori with me. If she cried and I wanted to pick her up, I had to ask a nurse to help because I wasn't in any condition to do it on my own. I was checked every hour, so I would just be nodding off and then they would take my blood pressure again. I don't think I have ever felt so tired, before or since, and that's including long Covid.

One nurse came in, took one look at me and then at Lori and said, 'Oh, this isn't your baby.' I was completely thrown, as she hadn't left my side. 'Yes, it is,' I said, sounding stronger than I felt. 'But you are dark, and she is not,' the nurse replied. 'What does your husband look like?' *And so it begins*, I thought. I had a vision of the future where I wouldn't be identified as Lori's mum. I was still on a drip, had a catheter fitted to empty my bladder, had lost a lot of blood and had a fresh Caesarean wound to contend with, so bonding with Lori wasn't easy. I did not need a nurse telling me my baby didn't look like me.

The next panic was the possibility that my bladder had been damaged during the birth, because my urine had an interesting pink tinge to it. They wanted to investigate.

CHAPTER 9

I wanted to get out of intensive care, where the lights shone too brightly every day and the same nurse who'd told me Lori wasn't mine also told me the reason my urine was red was not blood, it was the cranberry juice I was drinking. (I had done A level biology and knew this wasn't possible, but how come she didn't know it? I am still angry with this woman sixteen years later.) I escaped intensive care by paying for a private room, which also meant Martin could spend time with us both. This was not the time for penny pinching; I was in the middle of a medical mess-up. Also, working three jobs, I felt that I could afford it.

In the midst of it all, there was Lori. I know everybody thinks their babies are beautiful, but ours really was, with a sweet curl of hair in the middle of her forehead.

One of the doctors went through all the standard checks with Lori to make sure she was well. There was a tick box for each thing tested. The maximum any test could get was a 'satisfactory'. The doctor commented that satisfactory did not seem enough for the wonder that was Lori, and we both laughed at the silliness of that word that could not even begin to cover it. My baby was perfect; I couldn't take my eyes off her.

Lori was thriving, but I was not. They investigated the issue with my bladder by filling it with a tracer fluid and then looking for leaks, which, if found, would require an operation to fix. I had a lovely young female doctor who took me through the process of testing and told me there was bruising but no leak, which was a relief, until later when another doctor came to see me. He said he had the results, and it clearly stated there was a hole in my bladder, so they were booking me in for an operation. I was flabbergasted by this U-turn, as was the other doctor who had tested me. She went to the person who had written the report, only to find that they had made a mistake and had missed out a word, a rather crucial one. Instead of writing, 'there is not a leak,' they wrote, 'there is a leak'. Had the woman doctor not double checked this, I would have been back in surgery to have an invasive operation I didn't need. I don't remember her name, but I will never forget her professionalism, care and attention. Thank you, whoever you are!

My appetite returned soon after I had given birth, and both Lori and I found breastfeeding easy. I know that isn't always the case, but after the rough ride I had post-labour, I was glad something was going well. Although I was feeling better with each day, my Caesarean wound wasn't healing and, several weeks later, I was back in hospital with internal bleeding which had built up around my abdomen. It would take six months to fully heal.

There has been a lot of discussion recently around the higher risks associated around Black women giving birth. It is said that we are more likely to experience serious birth complications, our chances of dying in childbirth are almost four times higher than white women and we may suffer from a lack of appropriate care. I will never know if what

CHAPTER 9

happened to me is reflected in these statistics or whether I just got unlucky. A friend of mine had a similar experience with cholestasis and complications following a Caesarean, and she is white. What I do know is that a couple of times I was treated as if I had lost my mental capacity, or never had it in the first place. There was something dismissive about this, which I challenged.

I am not a passive person. In fact, I can be quite aggressive when I want information, and I am deeply offended when people assume I don't know anything. I think first-time mothers are often so vulnerable and silenced by the birth experience that they find it hard to speak up. As a scientist, I asked a lot of questions, fascinated by what was happening to me and how they were dealing with it. I wanted to learn from the bad parts of the experience.

There is no denying that a series of mistakes were made that could have been avoided and, had I wanted to, I could have taken legal advice – but to what end? I don't want to besmirch the medical profession, and most of the staff were lovely; they went above and beyond, even if there were also those few who didn't. It was torturous, and as I struggled through those early weeks, I missed out on precious moments when Lori was first born, but she was fine. More than fine. She was outstanding. I could go through anything just as long as she was OK. I felt as if I was holding my child aloft, wading through the medical quagmire of incompetence and illness.

STARCHILD

When she was older, I took Lori to Alton Towers, and we went on a water slide. She wasn't sure about it, but I encouraged her, saying we would go down together. As we flew out of the flume into the water, I realised it was deeper than I had thought; I went under the water, but I managed to hold Lori up high to keep her head dry. That sums up the story of her birth. She was my little miracle, and she still is – although at nearly 16 now, she'll kill me for saying it.

*

I have always been allergic to Brazil nuts, which I think is quite common. I discovered this when I was small, and my mum gave me a Brazil nut. Within minutes, my throat began to close, my eyes were puffy and I couldn't catch my breath. It was terrifying, and I remember Mum calling for an ambulance and talking to someone on the other end of the telephone. In the end, though, I wasn't taken to hospital, and clearly it didn't kill me.

Subsequently, I have avoided Brazil nuts which is pretty easy to do. But the severe allergy I developed after giving birth to Lori has proved much more of a challenge. Within weeks of her being born, my whole chest was raw with eczema and my asthma flared up. I got the same puffy eyes and I couldn't work out what was causing it, but I knew it was getting worse. It turned out to be an allergy to all forms of dairy.

CHAPTER 9

I am really strict when I eat out, but sometimes I ask the questions and don't get informed answers. Cafe and restaurant staff aren't always as clued up as they should be and this catches me out, usually meaning I end up vomiting violently soon after ingestion. Luckily, it happens once I have left. This is a marked improvement to how it used to affect me, which was my throat swelling up so I couldn't breathe.

I say I can't go near the stuff, but that's not strictly true. I can cook for other people and incorporate dairy in their food, I just can't ingest it. When Lori was a toddler, I gave her a little sippy cup of milk. I spilled some on my hand and absent-mindedly rubbed my eyes. I never made that mistake again.

Martin and I were travelling in Peru, and I had eaten something that was meant to be made with pecan nuts, but I reacted to it so I took an antihistamine. Later the same day, I must have had something else that triggered me, but the reaction was suppressed at first because of the tablet I had taken. It gradually got worse, and Martin called a doctor out who wasn't used to dealing with Black patients because he took one look at my swollen eyes and lips and asked, 'Does she normally look like this?' He brought out a bag of little vials and started to withdraw them one by one, looking at each one with a myopic stare before shaking his head, deliberating over which one to give me. This did not instil me with confidence. Martin and I looked at each other and

I pretended to feel much better, lest he give me something completely unsuitable.

I have been caught out on an aeroplane a few times and had to run to the loo, so I carry an EpiPen. The trials of travelling with an allergy!

Now we have 'Natasha's Law', legislation campaigned for by Tanya and Nadim, the inspirational parents of Natasha Ednan-Laperouse, who suffered a fatal allergic reaction after eating a packaged baguette that did not display allergen information. This change in law has had a hugely beneficial impact on my life and yet there are still many places that do not display the required allergen information, and when I ask to see the sheet, they can't find it. From my experience, it's harder in the USA, because without Natasha's Law in place there are cafes and restaurants that would rather not serve a person than risk legal action.

I was recently approached by the Natasha Allergy Research Foundation (NARF) and met Tanya and Nadim. They asked me to become one of their ambassadors and even though my dance card is rather full at the moment, I could not refuse. The work that they do is so important and life-changing, and I feel truly honoured to work beside them.

*

CHAPTER 9

Two days after I gave birth I checked my emails, much to Martin's bemusement. I had a message from the BBC saying they were working on an idea for a documentary – *Do We Really Need the Moon?* – and asking if I would be interested in presenting it. It was still speculative, hadn't been commissioned and wasn't necessarily going to come off, so without overthinking it I said yes. It seemed like madness to accept, but I could not help it. It was about the moon and I am a lunatic. After all, these things take a long time to get the green light and, in the meantime, I could enjoy my maternity leave. I just had one little commitment to fulfil for my UCL boss, Steve Miller: a series of lectures about science communication for the European Community. So, our first family trip, the three of us, was to go to Dubrovnik, a month after Lori was born. I was straight back on the horse.

I didn't have to go. Everyone would have understood, but it was a turning point for me. I was now a mother, something I had desperately longed for. I was also a career woman, something I did not want to give up. I thought I had found a way to make the two work together, but it was early days and I was tentative about my new plan. Going to Dubrovnik was a way of testing it. Steve was wonderfully supportive of a new mother taking her first wobbly steps back into the world of work.

I have always taken a strategic approach. Before Lori was born, I had it all mapped out: I would take maternity leave, settle her into the local nursery and then go back to work.

STARCHILD

But some branches of science can be cut-throat, and taking time off when you're approaching the peak of your career can be detrimental. There is inherent sexism, and this is an issue for female scientists.

With my three jobs, I was already juggling many balls even before adding a baby into the mix. When Lori arrived, something in me fundamentally changed. The morning after giving birth, I looked at her and my emotions were overwhelming, as was my need to protect her. It was like a wave of love rolling over me and it made me want to call my mum at 3am when I was breastfeeding, because I knew she would understand what I was feeling. I also knew I couldn't return to the life I had previously; I had to find a different way and I was ready for this evolution. The moon played a part in this because, when Lori was just four months old, I began filming the BBC documentary.

We travelled everywhere for the programme – up to Scotland, then Las Vegas, New York and Bermuda – and Martin and Lori came with me. Martin took this time as holiday leave from work. In hindsight, I think this was really hard on him, looking after a baby on location while the crew were focused on filming. Everyone has priorities, but they don't always align.

I have a photograph of me from that job. I am leaning on part of a stone circle up in Scotland. My eyes are bloodshot from the tiredness of night-time baby feeds. I am still recovering from the blood loss and protecting a wound

CHAPTER 9

which took a long time to heal. Martin and Lori are waiting in the car ready for me to finish so we can whizz back to the hotel and collapse. The sun is setting, and I am pretty sure we are done for the day when the crew whip out lights so we can continue filming. My heart sank lower than I thought it was possible to sink. I may have questioned my decision at that point.

But there was a method in my madness. I was going to leave Astrium and the world of industry and commit to my passion for science communication. I figured this way I could manage my own schedule, bring Lori with me to events and get the opportunity to be the mother I wanted to be. I was also lucky enough to be in a relationship with Martin, where we had two salaries coming into the household. Not everybody gets to choose. But, once I had made the decision and stepped away from research projects, I had a moment of doubt. I had worked so hard to become a scientist, but could I still define myself as such? There was only one way to find out, and that was to commit to the new direction I was heading.

My UCL fellowship had been extended far beyond what I had initially expected, stretching over four years, which enabled me to get up and running, and they were very happy with my work. It was an ideal collaboration. While they were keen for me to continue, they were also aware that they needed to give other people funding opportunities, so it felt like the right time to step back. This was a good point for me to expand Science Innovation into a more corporate fee-paying area, so I could use the money to support my charitable work. My relationship with UCL has developed over the years. It is now a less formal arrangement, but I still attend some events.

*

I wouldn't say I fell into television, but neither did I actively pursue it. It provided opportunities for my mission to make science more accessible, so I was intrigued by it, but I wasn't looking for fame or celebrity.

An invitation to appear on BBC News came through UCL. The physicist Professor Stephen Hawking, paralysed by motor neurone disease, was taking a parabolic flight to experience weightlessness, and this was one of the items on the programme. On my way to the studio, I was wondering how best I could describe what happens when the aircraft simulates zero-gravity, meaning the occupants can float freely for around 20 seconds. I wanted to get the story across rather than just explain the mechanics of it. I had an apple in my bag. Bingo! I could throw it in the air as a demonstration of the parabolic flight path, showing the path of the apple as it fell.

In the studio, I suggested this to the producers. They weren't used to live demonstrations, so they were a little

CHAPTER 9

uncomfortable with fruit being thrown around the news desk. They agreed I could do it on the condition that, if I dropped the apple, I would not go scrabbling for it. I think they had visions of a live broadcast of me crawling around on the floor with my bottom in the air. Luckily for them and me, it went well, and the apple really helped, so they started asking me to do more.

I was booked to go on BBC's *Newsnight* when Lori was just over a year old, so I took her with me. This time, my task was to talk about why the moon goes blood red during a total lunar eclipse. I have a demonstration that I use in schools to show that the sky is blue because, as the rainbow light from the sun passes through the atmosphere, the blue light is scattered, and this is what we see, but the red light passes through. I show this by mixing water with Dettol disinfectant and then shining a torch on it. The liquid turns blue. If you hold a piece of paper next to it, you can capture the red light that passes through. I reckoned this would appeal to Jeremy Paxman, although he seemed rather sceptical when I suggested it.

'You be the sun, Jeremy,' I said, giving him the torch, 'and I will recreate the earth's atmosphere with disinfectant and water.' For one frightening moment, I couldn't remember how much Dettol to pour in, but I winged it, and it worked.

Jeremy was impressed. OK, maybe he was mildly interested. Before we went live, I was getting my demo ready and Martin was holding Lori. Jeremy wandered into the green room where we were waiting and asked in slightly cutting tones, 'Whose baby is that?' Lori slowly turned her head, and stared straight back at him, scowling, as if to say, *Who are you talking to?* He scurried off and I thought, *Oh, I have a baby that can scare Jeremy Paxman.* She may want to put that on her CV now. Maybe she should also include the time we visited Downing Street when she was three years old, and she swerved David Cameron when he went in for a friendly hug. In fairness, Lori should have been more polite, because she wasn't technically invited. I found that if I was going somewhere and mentioned bringing my child, I would be told I couldn't, but if I just turned up with said child, there wasn't much they could do about it and generally Lori always received a warm welcome. Other than throw us both out, I suppose. Which wouldn't have been a great look for Number Ten.

I gave a talk at the Royal Institution when Lori was about 18 months old, and she was in my arms the entire time I was on stage. They invited me back to give the Christmas lecture, which was the biggest honour, and asked if I would bring Lori with me, which felt like a double bonus. The BBC transmit the talks every year, and I thought it would be such a lovely moment to treasure, of Lori and I on stage. Then, at the last minute, the BBC's plans changed, and they decided to go with someone else.

I was disappointed, not just because it was such a prestigious thing, but because Lori would have been part

CHAPTER 9

of it. It had been a secret tiny sore point for me ever since, but then last year they invited me again! This time, it was for their 200th anniversary. I did a talk there recently and showed a video clip of me holding her as a baby at the first lecture I had ever given there. I opened with that and then I welcomed Lori on stage, saying it was her turn to carry me! It was a lovely full-circle moment. At least this time around she didn't try to eat the fluffy cover off the microphone.

I also had no qualms about breastfeeding in public, and am possibly the only woman to have got her boobs out at both the Royal Institution and the Royal Society, on one occasion doing so on stage in the middle of an interview. Someone came up to me afterwards and asked if I had done that deliberately, which flummoxed me, as if I had fed Lori on stage to make a point. Gia Milinovich, presenter, writer and blogger, and wife of physicist Brian Cox, also came up and told me that when she saw me do that, it had made her cry. Which made me cry, too. The solidarity of women.

I think my confidence in taking Lori everywhere was partly down to my age. I had the bravado that comes with being an older mum and also knowing this was going to be my only child, so I didn't want to miss a moment. If I had to sit on a select committee then I wasn't going to be deprived of my daughter, so I brought her with me. They said she was the youngest member of the Select Committee they had ever had. I had a very rose-tinted view of these things, assuming it would work. Maybe the naysayers had a point, but I was going to try it anyway.

While this life Lori led with me made her hugely sociable and adaptable, as she got older, I was worried that she was happier in the company of adults than children. We would go to events together at places like the National Portrait Gallery, where she would eat delicious canapés and tell me what they tasted like. I couldn't eat them because of my dairy allergy. She would talk to grown-ups as equals, knowing that she had a right to be there.

I became a regular guest on the BBC's *Breakfast*, which was then filmed at Television Centre in White City, West London. I would turn up with Lori, and sometimes Carol Kirkwood, the kind weather presenter, would hold her while I was on air. When the production team was told they were moving up to the Salford studios, I was told that someone said, 'But what about Maggie? And Lori? That would be a much harder journey for them!'

It was really very simple. I loved spending time with Lori, and I also loved my work, so if I could combine the two then it was double happiness. Martin and I did book her a place at a local nursery, which was a great backup and meant that she could play with children her own age. It was useful when I had a lot of paperwork to do, but even then, I preferred to keep her with me.

I was lucky that I'd stepped back from industry to focus on the miracle of motherhood and the joy of science

CHAPTER 9

communication, and built a portfolio career that could accommodate this approach. I was my own boss, in charge of my day and no longer at the beck and call of industry or laboratories.

One memory stays in my mind. I was participating in a GCSE Science Live event and was on stage with Lori in the red sling I always carried her around in. I gave a talk about the universe and, at the end, a girl came up to me. She was in her teens, and she told me she had recently found out she was pregnant and thought her life was going to come to an end. She assumed she wouldn't be able to work and then she saw me on stage with Lori and she suddenly thought, *Yes I can. I can do this.* And she made me cry. Which made her cry. So we were both standing there crying and I told her that her life wasn't over, it was just beginning. I think of her often and I hope she is out there, living exactly the life she wanted with her child.

*

I have spent more time with Lori than many working mothers get to spend with their children, and I am truly grateful for that. There have been so many benefits for us both, but sometimes I worried I had taken it too far because, on the rare occasions I couldn't take her away with me, it would be hard for us both. As a small child, she would wander around the house looking for me and, after she had been in all the rooms, she would start again from the beginning, calling for me. I didn't know this at the time; Martin thought it would be better not to tell me, and he was right. This must have been hard on him, too.

Lori and I travelled everywhere together, especially across the UK, with my work. When travelling abroad, Martin came too. We went to so many places as a family that Martin put together a pinboard with a map on it showing where Lori had been in the first five years of her life. Each pin denoted a place she had visited, and he added a snapshot of us to represent each trip. She was a seasoned jet-setter at a very young age.

Then Lori started school. It was gut-wrenching for me to let her go, but I knew this was what she needed, even if it meant she couldn't come on the road with me as much. She looked so cute in her little uniform and was filled with excitement at the prospect of what was ahead. She loved her first week and, at the end of it, she said it had been great, but she wouldn't need to go back. She thought it was just for a week and then we would continue as life had been before.

The first thing I do when I am invited to travel somewhere for an event is to check the school holiday calendar. I would agree to short forays in term time if I had to, but I tried to organise my big work trips to coincide with the school holidays. Lori and I were invited out to Australia during one summer holiday for a science festival, and we

CHAPTER 9

stopped off in Sri Lanka on the way over. As a family we have been out to LA several times. I have three photos of Lori next to the Hollywood sign. Once when she was baby, again when she was a toddler and then recently as a sassy tween in big sunglasses.

There are many more travel adventures to come. I have a vague plan to buy a Land Rover through a sponsorship deal and deck it out for round-the-world travel. Lori and I would then make a documentary, *Lo and Mo Hit the Road*, charting our progress as we do a global tour, and give talks celebrating the wonders of our planets and our universe. Maybe across the plains of Africa, and I will bring my equipment so we can stargaze every night. It's the perfect location because of its remoteness and lack of pollution, so the night sky is breathtakingly clear, and you can see the expanse of the Milky Way.

Lori has been there for seminal moments that link me back to my childhood. I have been a fan of *The Clangers* for my entire life and have stated in many interviews that they are the reason I became a space scientist. As a result of this, when the series was revived, they invited me to appear in an episode as an astronaut. As it is generated by stop-frame animation, they made a little doll of me to visit the Clangers, only the second human to visit their world.

Lori came with me to the studio in Altrincham, just outside Manchester, where they film all the episodes, and at one point they put the Soup Dragon in my hand, which made me cry. 'Mum, it's just a puppet,' Lori said, bemused by my reaction, but the seven-year-old girl inside me was jumping up and down with glee.

Now Lori is older, she is much more independent; I can feel her pulling away from me, even though she also insists that she will never leave home. I have been a mother for 16 years and, while I don't think I have neglected other parts of my life, my initial thought is always of Lori. If something happens or if I'm offered a work project, I wonder how it is going to affect her. It echoes the relationship I had with Gracie through the first part of my life. Sometimes I accidentally call Lori 'Gracie' because the maternal response is so familiar. I took the position of protector for Gracie, and now I do the same for Lori.

Chapter 10
The Many Maggies

There are moments in life that feel so surreal they almost belong to another version of reality, like those parallel universes I wondered about earlier. One of such moments arrived when I discovered that two portraits of me now hang in the National Portrait Gallery.

The Gallery is a place I would wander through as a student, peering at the faces of writers, scientists, rebels and dreamers. These were the people who shaped history in small steps and giant leaps. I never imagined I would one day appear among them.

I was aware my photo was taken by the brilliant Simon Frederick, which is amazing in itself, but then came something even more astounding. My image was chosen for 'Work in Progress', the vast mural that greets visitors as they step into the Gallery's entrance hall – a sweeping celebration of 130 women, from the first century to the present day. To see myself in that setting, shoulder to shoulder with Ada Lovelace, Dame Emma Thompson, Queen Elizabeth I, Rosalind Franklin and so many remarkable figures, was almost too much to take in.

CHAPTER 10

Growing up, I rarely saw people who looked like me in science and certainly not gazing down from gallery walls. I often wondered where I belonged, or if I belonged at all. So to walk into that space and see myself reflected back felt like handing a message through time to my younger self: *You made it Maggie. And you were always meant to be here.* More importantly, I hope those portraits speak to others who might feel like outsiders, dreamers, or latecomers to their own potential. Let them look up and think, *If she can stand here, maybe I can too.* Because that is the true power of representation: not the portrait itself, but the spark it can ignite in someone else.

*

The toy brand Mattel made a Barbie doll of me. Every year they select six women from around the world as their role models, supporting Barbie's slogan, 'You Can Be Anything', and they make a doll in their image. They contacted me in 2023 and asked if I would be interested in being one of the women. Apart from the obvious flattery and honour, it also resonated with Lori and me because we used to adapt her Barbies, turning them into different characters. Sometimes we would even swap their heads round. Our goal was to create two dolls who represented us. And now here was the maker, Mattel, offering me the opportunity to turn into

Barbie. I was honoured to accept (and didn't confess that I occasionally beheaded their dolls).

I was tasked with sending through some head-and-shoulder and full-body photos of myself wearing the clothes I would like to be depicted in. As I got closer to the day I was to meet my doll in front of the cameras for the big reveal, I started to get worried: what if I didn't like it? What if it didn't look like me? How would I react, and would everything be caught on film for all to see? Would it be like one of those Christmas presents that you get and have to pretend is lovely?

I shouldn't have been concerned because Mattel were wonderful to work with and, when they revealed her, I was thrilled. She was a mini version of me wearing my space dress decorated in stars and moons, they even got the fine detail in my hair like my twists and purple highlights. The doll looked like me, but me on a really good day. The dolls always comes with a prop, so for me we chose a telescope. Even in doll form I am always looking for the stars. For a while, Barbie was the star attraction in my talks, and I was often asked if I could bring her with me to events. Everyone wanted to meet her. It was good timing because the movie had just come out, so it wasn't quite Barbie hysteria, but she was very popular. People would ask where they could buy one, but my Barbie is a one-off. She lives in my study next to the Barbie that Lori and I created of her.

CHAPTER 10

I never take my TV career for granted, nor do I have any expectation about where it will go. I learned this early on with my first telly job. I was asked to co-present an Open University documentary series in 2006 with Adam Hart-Davis, entitled *The Cosmos: A Beginner's Guide*. Other than a small blip when we went out to Chile to film at bHROS and weren't allowed access to the instrument because the room was occupied, it all went very well. I thought, *This is it, this is where my TV work takes off.* But of course it's not like that. That's one of the reasons I didn't pursue acting when I was younger: I didn't feel I could cope with the roller coaster of rejection and elation.

A few years later, after popping up on various news programmes, came the BBC documentary about the moon. This was followed by another documentary called *In Orbit: How Satellites Rule Our World*. At this point there was a lot of chat about me being the new face of space at the BBC, which I am still referred to from time to time.

As more TV work appeared, I thought it might be good to get an agent, but I had no idea how to find one. I was very fortunate because one found me. Vicki came to visit me soon after Lori was born, when I was about to embark on the moon documentary. She had had a similar upbringing to me involving lots of schools, and she was an older mum too. She had made the association between our experiences when listening to my *Desert Island Discs* interview on Radio 4. It was a match made in heaven. She is amazing

THE MANY MAGGIES

at all the detailed work and looking at contracts; stuff I find incredibly hard. I know that she has got my corner and, over the 15 years we have been together, we have grown close. I often talk to her about personal stuff as well as work.

CBeebies then asked me to present a series about stargazing, and I jumped at the chance. Lori was about five at the time, so it was a channel we loved and watched often, and I was always in my element speaking to kids in schools. It felt like a wonderful combination of the things I love most.

In lockdown, I also presented a show for CBBC, *Out Of This World*, and Lori joined me as my space apprentice which was such a fun experience. The production company making it had approached me, and I just happened to have a humungous green screen I had bought for Lori and me to use for TikTok and YouTube videos. I said we could create a little home studio, and they thought it was a brilliant idea. As we were in separate bubbles, we could not interact directly with the crew. When they arrived in the morning, Lori and I would make ourselves scarce while they set up the cameras in our kitchen dining area. Once this was done, they went outside to a gazebo they had set up in our back garden. Then Lori and I would come downstairs into the 'studio', taking directions from the crew via radios. It was a surreal experience.

As Lori and I are both dyslexic, neither of us enjoyed reading the script via autocue, but it was brilliant to conduct interviews over Zoom with amazing people from the space

CHAPTER 10

community like Tim Peake, Helen Sharman and Kathy Sullivan. Lori now has school friends whose younger siblings watched the series, but I'm not sure if she's proud or a little embarrassed by it now; she was happy about it at the time.

I was in the middle of filming for CBeebies in 2014 when I got a call about the iconic BBC programme *The Sky at Night,* the longest-running science programme in the world. I had appeared on it several times talking about various projects I was working on. I had never met its legendary presenter Sir Patrick Moore, although I knew of his reputation. He once wrote to the BBC claiming the demise of the corporation was because of its female producers, and he was known to be against immigration. I am not sure what he would have thought of me being considered as the presenter to follow in his footsteps.

Months before, I had been in touch with the controller of BBC Two and BBC Four, a brilliant woman called Kim Shillinglaw. Kim was one of the people who gave me my first big opportunity at the BBC on the Open University series, and I said I would love to be considered for *The Sky at Night* if the opportunity came up. As a child, I was allowed to stay up late and watch it; it was part of my psyche.

After Patrick's death, there was some talk about cancelling the programme, which caused an uproar. Then it went very quiet, so I thought that was the end – until I got that call. They told me the show was going to continue, and asked if I would like to co-present it with the great

THE MANY MAGGIES

astrophysicist Chris Lintott. I was speechless, although not for long. Another dream realised.

Filming my first episode, I was like a rabbit caught in the headlights. I didn't know what was going on and I was terrified. Patrick's shoes were iconic ones to fill. I knew I couldn't do what he did, so I did it my way and I am glad I made that decision early on. Some people say we have several things in common, from our fast talking to our boundless enthusiasm, but I think what really shows is our shared passion for space.

While I was very comfortable in front of the camera, I had never done anything as scripted as *The Sky at Night* before. I persevered with learning the lines, but my dyslexia tripped me up. Some of my early directors wanted me to repeat everything word for word and, each time we did another take, it would come out differently. I found it hard. I didn't want to let the crew down, but it crushed my natural exuberance because I would be thinking more about not forgetting lines than about what was being said. Now, some 12 years later, I'm given bullet points to talk around, which suits me much better. The subjects we have covered and the people I have met during this time have been, and continue to be, an absolute joy.

I got to see another side of television when I was asked to be a science consultant on the BBC One drama series *Paradox*. The series was about a chap who was getting images from a satellite, but they turned out to be of events before

CHAPTER 10

they happened on Earth. It was like a superpower he could use to prevent disaster.

I met up with the amazing actor Emun Elliott several times. He was playing the role of space scientist, so wanted me to share with him the realities of the industry and show how a scientist behaves. He watched my every move very carefully, even how I would hold a pen and draw diagrams on paper to describe principles of physics. As a great actor, he imitated my movements precisely. I am not sure if I was the best example, but it was interesting to be observed in that way, and I'll take any excuse to talk about science.

As well as briefing the actors, I was also involved in the storylines. There was talk of a story about satellites seeing car number plates, but I had to explain that it wouldn't be possible to see number plates from space using a commercial satellite; only military satellites are powerful enough to see something that small, and also the orientation of the number plate would stop it from being observed from above. Of course, there was some poetic licence and compromise involved, but everyone was keen to get the essence of the concept right. They sent me scripts to review, looking at the logistics and dialogue and providing a view on the scientific rigour of it all and I visited the set several times, in awe of the enormous production crew, including the catering and wardrobe. We never got this in factual programming!

*

THE MANY MAGGIES

It was a fascinating journey for me and one I got to repeat when Sky One asked me to consult on one of their dramas, *Intergalactic*, about women criminals in space. It was shot in Manchester, and they built a mock-up of a spacecraft that they asked me to help design. Never in my wildest dreams would I have expected to do that. I have a video of Lori running through the corridors of the spaceship while we pretended that we were in space.

I loved this project, because it was sci-fi focused on a group of women. One of the characters had plaits, longer than mine, and they were mechanised so she could attack people with them, which I thought was a brilliant idea. I initially had a slight issue with the title because 'intergalactic' implies traveling from our galaxy to another one. As this distance is 3 million to 100 million light years, it seemed like a real challenge. To reach our closest star within our galaxy using current technology would take around 76,000 years. The production team had already looked into this and adopted a theory of the Mexican theoretical physicist Miguel Alcubierre. He proposed that rather than travelling faster than the speed of light (Einstein's theory of special relativity indicated that nothing can travel faster than light), you could squish space and time in front of you and elongate it behind you. So the craft does not move faster than light, but by warping spacetime you could travel vast distances in an instant.

CHAPTER 10

To do this required excessive amounts of energy, which the programme solved by introducing a new element via a comet that had landed on Earth and provided a substance that released the vast amounts of power needed for space travel. In the show this substance was initially called Blue Mercury. The production changed the colour to amber after I attended the premiere of the film *Captain Marvel*, which featured a blue substance with similar miraculous properties. Although they had been conceived completely independently, I thought there was a bit too much of an overlap.

I enjoy science fiction most when it takes current ideas and extrapolates from them so you can see the basis of truth: it doesn't require a complete leap of imagination and stays within the realms of possibility. Taking something that's familiar, plausible or that you could see happening in the future, makes science fiction more powerful.

Here's the thing about being in front of the camera: you don't have any control over what opportunities come your way until you become a big star and, even then, you can be dashed by a turkey of a show. I love doing TV both in front and behind the camera, but I am now wise to its fickle nature and accepting of the challenges it presents. I don't dwell on the things that don't happen, otherwise they would eat away at me. I take each new opportunity as a positive, but I am not reliant on it financially or emotionally, and I know it could all end tomorrow.

*

THE MANY MAGGIES

My father always told me not to worry if something didn't happen the way I wanted it to go. With my hyperfocus it is easy to get caught up in ideas and find it hard to put them down. Even today I can hear my father's voice in my head: *Don't worry, Mag, this or something better will come along.* I find it such a reassuring way to look to the future and it stops me from fixating. And looking back so far, I think that he was right – I am generally happy with my lot in life.

That said, I have a competitive spirit. I do lots of panel and game shows now which promote healthy competition, and I love them. I've realised through doing these things that my processing is quite slow, so I need a little time to think. Being on these shows has taught me more about my capabilities. My recent dyslexic assessment has also confirmed my view on this. Being tested for dyslexia was not a decision I took lightly, the tests are very expensive and it wasn't necessarily essential for me to know as an adult. However, being diagnosed flagged up my slower processing speed, which is the rate at which the brain takes in, analyses and responds to information. I sometimes struggle with tasks requiring quick responses or the ability to handle large amounts of information at once. This affects everything from reading and writing to following instructions and completing timed tests. Knowing this about myself has meant that I've been able to curate my work more efficiently, working with my brain and not against it, and making sure I set up the right conditions to tackle the work I need to do.

CHAPTER 10

Slow processing speed is a common characteristic of dyslexia, and it does not reflect a person's intelligence but rather their unique brain functioning. Also, although I am slower at some tasks, I am lightning quick in other areas; from a question on stage about the universe to a witty retort on a comedy show. When Lori found out that I was considering going for the test, her reaction was 'Don't do it, Mum! What if you're not dyslexic, the game could be up!' Luckily, I was!

Years before my formal diagnosis, I mentioned in passing to a journalist in a pub that I thought I was dyslexic. I was being interviewed for a newspaper, and I was drinking a cider which may or may not have had something to do with being so frank. I had never shared this before or spent much time thinking about it, and afterwards I wondered if I should have said anything.

The piece went out and there was a response to the dyslexia revelation which made me realise how important it was to speak up and raise awareness. I began to include it in my public speeches and talked about 'suffering' from dyslexia until I was contacted by Kate Griggs, the brilliant woman who set up Made By Dyslexia to support her son's journey through the challenge of the school system. She invited me to be an ambassador. I looked at their website and realised my terminology was all wrong; I wasn't suffering, I was gifted with dyslexia.

THE MANY MAGGIES

I saw that all the Maggieisms I assumed were my own oddities were recognised and celebrated by the organisation. Dyslexics are storytellers. We are explorers. We think outside the box. We are curious. We are problem-solvers. We are communicators. We have stamina and resilience. It was crystal clear to me that I had steered my career in a direction where I was working to my strengths, partly guided by my dyslexic qualities. This was positive, not negative.

Seeing myself reflected on the website was a gamechanger. It was cathartic and liberating, as if I could take deeper breaths. Suddenly the way I thought and behaved made total sense to me. I read a quote from the wonderful dyslexia researcher Dr Helen Taylor, who said that dyslexics are often the pioneers, the people who push boundaries. I see this in the work I now do with Made By Dyslexia. I like to push those boundaries and do things differently, and Made By Dyslexia enabled me to understand why.

Isaac Newton is believed to have been dyslexic. Many of the game-changers in science were or were purported to have dyslexic or neurodiverse thinking, including Galileo, Leonardo da Vinci, the Wright Brothers, Einstein and also Stephen Hawking. Dyslexics don't just think outside the box, they think off the planet and beyond.

I meet people who meandered through the education system and for whom life only began once they were out of it. These early experiences as undiagnosed dyslexics in

CHAPTER 10

a classroom gave us resilience. We learned from failure and are well-versed in the struggle, but some people don't hit stumbling blocks until later in life and are less equipped to deal with them.

I am so glad that my understanding of dyslexia came around the time Lori was diagnosed with it. Martin and I were sure that she was, but it is not something you can formally investigate until a child is eight years old. We once went into her primary school for parents' evening, knowing that the teachers would not be able to comment until she was older and as soon as we mentioned it they said, yep pretty sure she is. *Oh*, we thought, *so no grey area there*.

For Lori, the knowledge has been a superpower. I love this attitude to it, and seeing her approach has encouraged me to celebrate my own dyslexia. I spent so long masking it and using coping mechanisms to minimise what could be perceived as 'strangeness' in my behaviour to fit in. I talk about it a lot in my work and share both Lori's and my personal experience of it, with her blessing. She impresses me so much.

*

The first time I was called a role model, it made me feel uncomfortable. I thought of my flaws – like my untidiness, tardiness and disorganisation – that felt the opposite of an inspirational person. I couldn't be a role model because those are people who get everything right. It was a classic case of imposter syndrome. Then I worked out that maybe the point was to be imperfect. If I go into a school to talk to students and I have a bit of jam on my top or look a bit scruffy and I say I am a space scientist, then it may feel more attainable to them. They are more likely to think, *Oh she's normal, she's just like me, maybe I can do stuff too.* I want to explode presumptions of class, gender and race and promise them if I can, they can, as early in their lives as possible. And I tell them we are all role models because we all have something to share, and we don't have to be perfect. In fact, I think not being so makes us more effective.

The nature of my media and television work means I occasionally get to meet my own role models, people who have unwittingly played a significant part in my life. Like Sir Michael Caine, an actor I am a huge fan of. I was invited to give a talk at a Microsoft event. I was on the bill before Sir Michael, who was the headline speaker, and I guessed the audience would be politely watching me but counting down to the main event. So, I leaned into this and turned my talk into a warm-up act before his entrance, looking at the overlap between his science-fiction films and my career in science. It worked. As I came off stage, he was there in the wings.

'I have just been listening to you,' he said appreciatively.

'I have listened to you all my life!' I responded. I was

CHAPTER 10

utterly starstruck, but not too overwhelmed to ask for a photo together. He was so charming.

More recently, I met Tom Hanks, and found out he is a fellow lunatic and a thoroughly lovely man. He had narrated an immersive exhibition called *Moonwalkers*, and I was invited to its opening. I brought a copy of my book about the moon with me and asked one of the team if they would pass it on to Tom on my behalf. They said that they could do better than that and ushered me into another room and there, in the middle of it, was Tom himself. I had not been expecting that, nor did I think we would be introduced, but to my delight we shook hands and I gave him the book.

'I love the moon,' he said excitedly, 'and when we meet again, this book will be dogeared, with the corners of pages turned down, and I will have lots of questions for you to answer.' Oh, Tom.

Sandi Toksvig is another of my heroes. I have known Sandi for several years and taken part in her radio and TV shows. Her fascination with space started very young, when as a child she was at mission control during the moon landing because her father was a diplomat, and she held on to the hand of Neil Armstrong's secretary. The woman was so worried about him, and Sandi wanted to reassure her. What a magical moment.

THE MANY MAGGIES

Then there's Sir David Attenborough.[6] Unlike the fleeting moments with Michael and Tom, I sat next to David at a dinner. It was a Reith Lectures event organised by the BBC, which I had been to before (I met Stephen Hawking at the first one). David and I got into a debate about life forms on other planets, and he was firm that life needs water.

'Oh no, Sir David,' I said, disagreeing with our national treasure, 'that's just life as we know it. Maybe there's life we don't even recognise. Like little Martians who hide away from our prying eyes?'

I told him I wanted to go into space and investigate this for myself, then I could tell him what was really happening up there.

'Maggie,' Sir David, said looking me straight in the eye, 'you are a real star, and we need you here on Earth, so don't go out there.'

Be still my beating heart.

*

6. Look, I know how this sounds. I couldn't decide if I should name-drop throughout the book or save it all up for one big session of showing off. I decided, on balance, it may be easier to stomach in one go. Besides, these are my role models, and it's important to say that, after meeting them all, I can firmly state that they still are. Very much so.

CHAPTER 10

Stars go through life cycles. A star is born in a cloud of dust and gas called a nebula. It heats up, growing to a critical mass where it starts to shine brightly before, depending on its size, it will die or fade into something different. They can live for a few million and up to trillions of years. Our local star is the sun, which we see as constant and inactive, but it still goes through its own life cycle, albeit over around nine billion years. These phases fascinate me, both in space and in life. The cosmos isn't static, and neither are we.

I know some people grow incrementally, but I don't think that's my approach. I stay in one skin for a while and then I shed it completely, to become a new Maggie. As I write, I am going through a re-evaluation, a new part of my life cycle. I have lived a consistent life for the last 15 years and now it is time to change, like the Doctor going through a regeneration. I grow and expand through redefinition, but it isn't a continuous journey, it is in quantum leaps: abrupt transitions that can be a shock for those close to me. I am not sure where it will take me. I look at my mum who is roughly 30 years older than me and I think, if I am really lucky, I will get to her age, and I don't want to waste a single minute. In this next stage of my life, what do I want to do and who do I want to be?

Looking at my mum, I see that she had been a chameleon too. From her early life in Nigeria as a very young mother to my half-sister, to coming to the UK and finding her place. First with my father, then in the vicarage and for a while independently. I see myself following in her footsteps. When I was in Hasting, she used to appear on television on a program called *House Party*, like an early version of *Loose Women*. She also spent many years working as a youth and community worker, inspiring kids and running after school and weekend activities to keep them off the streets.

We are all made up of so many parts of ourselves. I am Maggie the mother, Maggie the scientist, Maggie the Black woman, Maggie the daughter, Maggie the child, Maggie the wife, Maggie the presenter, Maggie the evolver, Maggie the astronaut, Maggie the unmanageable. So many Maggies. I have transformed at various stages in my life, and I have never been afraid; I have relished it. I can stay static for a while, but it is not my natural state. Change is in my core, it is my oxygen, my security. I know many people who run away from it, but I run towards it, arms outstretched, embracing the unfamiliar.

I am no longer Maggie the wife. Martin and I have decided to separate. It's hard to write about this coherently because we are still in the middle of it all. Like being in a big cloud that is only possible to describe once you are out of it and can look back and see its shape clearly. Also, this is not just happening to me, it is happening to Martin and Lori too, so it would be unfair to unravel parts of this current story that are not mine alone to tell.

I take responsibility for the break-up, but I don't take the blame. This is something Martin and I carry between

CHAPTER 10

us. We have drifted apart and found ourselves heading in different directions. I'm not even sure how and when this happened. And while change is my natural state and I am ready to throw all the cards up in the air, it is the opposite for Martin, and I understand this takes longer for him to adjust to. We have been in counselling, not to try to repair our marriage, but to help us communicate through the break-up and make it as amicable and easy as possible. Lori is our priority and beyond that we want to continue to be kind and supportive of each other.

I can make hard decisions, those that I know could hurt people, but that need to be made. I have a steely determination that I rely on in difficult times. Maybe this has been nurtured from my early experiences, but I think it is a fundamental part of me, no matter my upbringing, and I have always had a strong streak of stubbornness that has mostly served me well. Stubbornness was seen as negative, but I came to understand it as a protective strength, a compass that kept me true to myself.

Being stubborn was used as an insult when I was growing up – it was seen as a negative, but it's not. It feels protective, like a boundary. I have been able to stand up for myself when it matters to me.

Stubborn optimism is a wonderful term that a brilliant woman, Costa Rican diplomat Christiana Figueres mentions in a TED Talk. She said stubborn optimism was something she learned from her father, who had been the

THE MANY MAGGIES

president of Costa Rica. Optimism is great because it gets you up in the morning and encourages you to do things, but it is a fair-weather companion. It is easy to lose heart and give up, so what we also need is stubbornness, which can be brutal but necessary. The two together make excellent bedfellows. Optimism makes you aim high, but stubbornness is what is going to get you there.

It's the ultimate combination for survival, and it sums me up. I can pinpoint the moments in my life when it has worked for me. During change and uncertainty we should look for opportunity. It's always there if we are open to it and brave enough to step forward.

*

A while ago, I applied for a job as a professor of public engagement of science. It was a role I really wanted, as it felt directly aligned to the career I have built in science communication over the last 20 years. Plus, with my change in marital status, it would have been good to know I could rely on a regular income to support Lori and me, rather than solely relying on freelance work. The application process was convoluted for various reasons, including me needing to be affiliated to a university. It took a lot of time, energy and help from a range of wonderful people and, in brief, I didn't get the job. I share this, though, for two reasons.

CHAPTER 10

Firstly, it's important to talk about the failures, the things that don't work out. Research does not always come out the way we expect it to. But that is the nature of science and the scientific process. We postulate an idea, a theory, we look into it and assess: has it been done before? Is there merit in further investigation? We work out what to expect from the research. We then carry out the experimentation and see how it differs from what we predicted. Most often the prediction and the experimentation don't coincide, but this is not a failure, it is an opportunity to learn. The different direction the experimentation takes us gives us more information and tells us which assumptions are incorrect, so we postulate a modified idea and start the process again. Each deviance or perceived dead end takes us closer to the truth. So, with time, for all of the many failures in my life, I like think that I can learn from this. There is no shame in really wanting something and not being able to achieve it. Secondly, applying for the job made me realise more about myself.

As a child, I was desperate to be an academic and I spent my entire education focused on achieving it, until I got to the point where I could be, and I realised that wasn't what I wanted at all, and it did not fit my skill set. Yet, I went for this job and was likely unsuccessful partly because I am not an academic. Nor do I want to be. I like doing things differently and I celebrate the way my career has crossed boundaries and led me to media and high-profile opportunities.

I think there is still a snobbery around being a science communicator and not an academic. People often ask me how I juggle this role with my research, and the answer is I don't. As a researcher I was reasonable, but not brilliant. As a science communicator, I think I can really make a difference, and that is more important to me.

Knowing your strengths and weaknesses means you can steer your way into the things that make you happy, fulfilled and successful. A few years ago, Martin introduced me to a Japanese concept called Ikigai, the art of life fulfilment. It's a Venn diagram that shows the meaning of life, with four circles representing what you love, what you are good at, what the world needs and what you can get paid for. If you can combine all four in your life and work, that bit in the middle where they all overlap, that is Ikigai.

I took part in an event at Lori's school recently. It was career speed dating, which is quite intense because my fellow advisors (other parents from different disciplines) and I were seeing around 180 pupils in a day and each pupil for just three minutes. What pearls of wisdom can you give in that time? I turned to Ikigai. I think it is a message for everybody. Do what you love and you will never work a day in your life.

I wouldn't have chosen to set up my own company, but it has proved to be one of the best decisions I ever made. Science Innovation recently celebrated its 21st anniversary, and in that time it has grown enormously, built a healthy

CHAPTER 10

turnover and reached hundreds of thousands of people. I am hugely proud of what I have achieved, sharing my knowledge and experience, encouraging debate and aspirations and still learning about others and myself in the process.

One of the most enjoyable and rewarding parts of it is meeting people after events. I feel so honoured to have these connections with strangers and to hear their stories. I have been doing this for a long time, and it never wears thin. I can't think of another job that would have given me as much or opened so many doors.

*

There is a very famous photo of Earth from 1966 called 'Earthrise', taken by Apollo 8, one of the space missions that didn't land on the moon but orbited it. The astronauts saw a glorious image of our own planet rising beyond the moon and one of them said, 'Quick, pass me the camera, the one with the colour film. Everybody at home needs to see this.' They were right, everybody did need to see it, and they were bowled over. They saw our planet in a different light, and it was a total game-changer. It looked small and vulnerable in the darkness of space. In fact, many people believe this image spawned the environmental movement. Suddenly we realised that we needed to look after our planet better, although I am not sure how successful we have been at that.

THE MANY MAGGIES

Recently, I became an ambassador to a world-changing organisation called the International Rescue Committee (IRC). It's one of the projects most close to my heart. The IRC aims to help people across the globe, but especially those in the worst affected areas, whether they have been devastated by war, climate, drug cartels or other reasons. Every year, they come up with a watchlist of the 20 countries most affected by humanitarian emergencies. The aim is to link the people (or clients, as they call them) in these areas with the help and support that they need, firstly to survive and then to thrive. I was gobsmacked when I was invited to be an ambassador. I was joining a who's who of talented celebrities such as George Clooney, Natalie Portman, Rami Malek and Sir Patrick Stewart. Or, as I like to say, Maggie and the great and the good. I was so honoured to be asked to join such an illustrious group of meteoric high-profile people, but I started to think, *What do I have to offer?*

I met up with their president and CEO David Miliband when I was first invited to join, and he told me about the history of the organisation. It was set up in 1933 at the suggestion of Albert Einstein to raise awareness of the plight of refugees fleeing Nazi Germany. Einstein's wise recommendation for the organisation was a strong commitment to research, evaluation and learning, and they continue to operate under this premise today by challenging assumptions, exploring outcomes and using evidence.

CHAPTER 10

At its core, there is a scientific focus which really appeals to me, and they did not have any scientists as ambassadors, so when they approached me to ask if I would participate it was an instant and resounding yes. A couple of years ago, I went to a school to talk to refugee children from Ukraine about the challenges and cultural differences they faced as they settled in, and the *Guardian* came along too. The IRC has people who come in and speak to the refugee students from whichever country, and they also give the school an understanding of where these children have come from and what they have faced. One of the future trips I have planned with them is to Nigeria to encourage more girls into STEM subjects.

The work the committee does is amazing and empowering, covering so many global issues. For example, they may step in and help farmers when a disaster hits, perhaps with seeds set aside that they can give them. They can also help people work out what they need, and give money to facilitate this.

I filmed a piece for them about malnutrition in babies and how this has been measured in the past through tests and assessments before they could work out what formula to give them. The IRC were suggesting this could be simplified to choose one formula that works for all kids across the board. This is what the organisation are so good at: looking at something that has always been done a certain way and – without being patronising – proposing a different approach. They offer solutions and ways to update processes that make them more efficient. I often attend their fundraisers, no matter what I am doing.

As a scientist and an engineer, I want to make a difference – that's why many of us scientists do the jobs we do – but it isn't always obvious how we can make that difference. We need to know about the problems in order to solve them. I have been researching where I can add value because, as great as it is to be invited to take part in events, visits and communication, I am keen to come back to the committee with my own ideas. I have been working on one with the Royal Society which I am hopeful we will get off the ground, looking at the global research labs and scientists the IRC have at their disposal. It is something we can all contribute to.

*

I never imagined that my work would win me awards, influential listings, honorary doctorates and presidencies, and I feel incredibly humbled as the recipient of these wonderful acknowledgements. One of the positions I was most proud to be offered was the chancellorship of the University of Leicester, following on from the honorary doctorate they gave me. I was approached by their president and vice chancellor, Professor Nishan Canagarajah, a wonderful man and

CHAPTER 10

the person who does all the actual work. The chancellor role is more ceremonial, a smile-and-wave job as I call it, but I love being at graduation ceremonies and seeing a wide range of people from different cultures, of different ages and at various stages in their lives using the power of education to transform.

The University of Leicester was a very good fit for me, partly because Leicester is the first 'plural city' in the UK, which means no one ethnic group is dominant in the demographic. Every culture collaborates and aims to live in harmony. Plus, the university has a fantastic space department. They have been involved in many of the missions since the UK got into space. Their professor of space physics, Ken Pounds, worked with the Americans early in his career. When the Americans were detonating nuclear explosions in the atmosphere, he was building sensors to detect them.

As well as graduation ceremonies, I give talks at the university and I'm active on their social media, taking part in TikTok trends, adapted by the brilliant Carla Creary from the communications team. She will spot something online and mention it to me; for one video I had to dance in my chancellor's robes with other members of staff. I love the ingenuity and informality to this part of my role, as well as being one of the spokespeople, supporting the work the faculty are involved with.

One of the other areas where the University of Leicester excels is clinical trials. Leicester is a highly diverse city so local clinical trials fare better than most. For many of these trials, the clinicians need to broaden the demographic of people who take part in these studies as it isn't currently very diverse. I have been involved with clinical trials, and I know how crucial they are. I was filmed participating in one of the trials, advertising to the public the importance of diversity for these studies and encouraging them to take part. The study was to look at the correlation between diabetes and heart problems. To participate I had a heart MOT, and they did various tests looking at blood flow and efficiency of pumping.

Last year I went to LA with some of the team to visit alumni and encourage them to help support the future of the university. We also visited Mattel while we were there, and NASA's Jet Propulsion Laboratory. I may be a figurehead, but I put time and effort into adding real value to the university.

There is no better way to spend time than hanging out with inspirational and passionate young people; it energises me. I was on the judging panel for the National Science + Engineering Competition and absolutely loved it, meeting so many teenagers in the process. The idea behind this and other similar initiatives, like the CREST Awards, is to encourage students to participate in simple research. When I was at school I entered Young Engineer of Britain, which was also run by the British Science Association, so I was thrilled to become a judge for it.

CHAPTER 10

The awards receive thousands of applications, so it is a long and involved process to go through them all. What always surprised me was the quality of the science I was judging from 12- to 16-year-olds, including a version of genome sequencing and a simple but effective baby birth monitor for the developing world, which enables pregnant mothers to time their contractions. This work was born out of a passion for a subject, the desire to make the world a better place and a spark of entrepreneurial spirit: a powerful, unbeatable combination.

Judging this competition, coupled with the work I do in schools, makes me certain that our future is in good, safe hands. The younger generations seem to be aware of the challenges and want to make a difference.

*

When I was given an MBE in 2009 for my services to science education, I nearly fell off my chair. So, imagine my shock when I opened a letter a couple of years ago that asking me if I would be happy to be considered for a damehood. I literally couldn't believe my eyes and I had to keep rereading to make sure I had understood it correctly. There was an online form to fill in to formally accept the honour – which I thought I had done, until I got a call a few weeks later to ask if I was interested in a damehood because they had not heard back from me, and it was past the deadline. I couldn't believe I had fluffed the response. Well, actually I can believe it – who am I kidding?

This time around I was given the honour for services to science education and diversity, which felt like a double win and meant everything to me. There aren't any commitments to fulfil with the title, other than continuing to maintain what I am doing and going to the Dame Commander gatherings. (I certainly don't march through Tesco saying, 'Make way, Dame Maggie coming through!' – although it's tempting.) It does bring more attention, and I am asked to get involved with a lot of initiatives and charities.

I am recognised in public occasionally, which is nice because it's mainly positive, but I can also get on the bus and nobody knows who I am, so I think I have the best of both worlds. Sometimes I do have to pinch myself. I am proud of how much I have achieved, sometimes against the odds. As a Black woman, I have faced a range of challenges. Believing I was worthless and giving up was not an option for me. I thank my mum, who showed me how to be a strong woman, and my father for inspiring a can-do attitude in me. I carry elements of both my parents within me. They are just two people who were ill-suited and one of them decided to leave.

Mum is a strong, powerful woman, and age has not diminished her. She will enter a cafe, and on realising there is no seating available for us, she will get us to move chairs

CHAPTER 10

around and invite people to reposition. It's an energy that can feel a bit overwhelming but allows her to make changes in the world, just as she has always done. Growing up, I saw her making waves as one of the few Black women on TV, and that pioneering spirit lives on in her but also in me. It is a spirit that has shaped my own determination to claim space in the world.

When I received my damehood, I invited her to come with Lori and Martin to be with me on my special day. We had a picture taken of Mum, Lori and myself, ladies together in our glad rags. I love the picture as it is three generations of women, each making their way through the world, each taking a different route.

Mum and I don't talk about the past much. It fires up so many emotions, it's too dangerous to step back into. But in writing this autobiography, I am of course dredging up that past.

I asked her to dig out some photos of my childhood and she did, but to my surprise she also found letters that I had written to her between the ages of 14 to 21. There aren't many, but I thought that we were completely estranged during this time. It goes to show that there are many emotional aberrations in the retrospectroscope. I think that my remorse about leaving made me feel that we were out of contact for years, whereas we were in intermittent contact via letters. They are a little painful to read. They start with 'My Dearest Mum' or 'Dearest Mummy' and end 'from a daughter that loves you very, very much'. I think my heart was in the right place, and the letters show how proud I was of myself for my educational achievements against the odds. But they are also very nerdy and cringey. In them I ask her to look after herself and tell her how much I love her and miss her. It seems that we also met up occasionally, but this is captured in the text rather than my memories. But the fact is that even after the harsh severing of our relationship, love still flourished.

Moments like this are hard, but it does feel like reconciliation. Our family was fractured from the moment my parents split up, possibly from before, and it has remained so in different ways and for different reasons over the years. We tread carefully, like passing through the minefield of our past. Unfortunately, we do not have the sophistication of ground-penetrating radar or nuclear quadrupole resonance to guide us. We are just using rudimentary equipment like a long stick to gain passage. If one of us mentions the wrong thing from the past it can cause a knee-jerk reaction from someone else. It means that we are not very connected because there are too many potential points of ignition, and the slightest trigger can set them off. I am hoping this book won't be one of them.

We meet up very occasionally, and I wish we could do it more, but there are very few events that will bring us all together other than weddings and funerals. Maybe we all feel safer being apart, in our own independent lives.

*

CHAPTER 10

I am within touching distance of 60. In many ways I feel no different from the small girl with big plans. I can tap into her energy, her zest and joy for life. My spirit is still vibrant even if my body creaks and groans due to the ageing process. The last ten years have been physically challenging for me, and I have dealt with various health issues that remind me how precious life is.

I have had Covid twice. The first time it was like an annoying cold that I was able to brush off with ease, a minor irritation. The second time I was very poorly, the sort of ill where you can't get out bed, and this turned into long Covid. Unbeknownst to me at the time, this coincided with perimenopause. After having Lori I had the progesterone implant put back in – a tiny pin which sits under the skin of the upper arm – and had not had a period for years, which was bliss after the terrible time I'd had with endometriosis. The downside was that I had no idea I was going into the menopause.

I wasn't sure where Covid ended and perimenopause began; it was just a prolonged period of utter exhaustion. I tried not to let it affect my work, but as soon as I was at home, I just wanted lie down in bed, which wasn't fair on Lori. I hate taking any form of medication, but after reading the benefits around HRT, particularly the potential reduced risk of Alzheimer's, I gave it a whirl and have been on it for a couple of years. It has made a big difference, particularly to my energy levels.

I am also on a countdown for a knee replacement, as my prognosis in this area is bleak. After a couple of MRI scans, it was clear that the degeneration was rapid. At their arthritic worst, I had mobility issues and had to use a wheelchair, which gave me an insight into the life of a wheelchair user, from the impracticalities of the world around them to the way they can become invisible. People literally spoke over the top of my head as if I wasn't there. It was a sobering experience, and it has made me more aware of the problems less able-bodied people face.

I was also advised to lose weight to take the pressure off my knees, which I have done, and it has worked well so far. The first day back in the gym for a long time, I bounded over to the bench press and loaded the bar up with the weights I used to be able to lift. I tried. Nothing. I took some weights off. I still couldn't shift it. My muscle tone was non-existent. In midlife, strength training is as important as cardio, and a combination of the two is ideal, but preserving muscle mass is key. Mine was a disaster, so I have been focusing on this.

Losing strength is a frightening feeling. I used to run like the wind, lift heavy objects and have excellent overall general fitness, particularly when I was cycling to and from university every day. With the weight loss and improvement in my health, I am determined to be in much better condition as I go into my next decade.

*

CHAPTER 10

I believe in happily ever after. I want good to triumph. Maybe this is why I have a terrible habit of trying to make people's dreams come true. I do it a lot with Lori. She may mention being interested in something and I am already working out how to make it happen. I am not sure if that is particularly healthy.

For a while, Lori had a fixation with sloths, and we collected soft toy versions. Around this time, I was invited to attend a literature festival in Dubai, and I arranged for Lori to come with me so we could visit Green Planet where you can meet real sloths. This was going to be a dream come true. It was a lovely moment for us both, but soon after her interest waned, and she moved on to something else. But it was too late for me. After meeting them, I became a sloth girl. Now the toys are in my bedroom and I prop them up, make sure they are comfortable and go to sleep each night with 21 sloths looking at me.

Achieving your dreams is one of the central themes in my lectures and talks. We all need a goal, something to aspire to and aim for, no matter who we are and what we have already done in our lives. I feel privileged to be able to share this message and connect directly with the audience.

I try and tailor my speech for each audience. Talking to children is incredibly enjoyable, particularly when I feel the moment I have captured their attention and curiosity. I love taking part in events for International Women's Day, as I love discussing the challenges we share and how far we have moved the dial. We talk about how much we can achieve if we work together and I say, 'Come on ladies, let's take over the world!' I am not aiming to incite a riot, maybe just a small bloodless coup.

I was invited on to a Saga cruise recently to give a talk and take part in a Q&A. The audience demographic was in their 70s, so my student talk about having their entire lives ahead of them wasn't appropriate, but my core message remained the same. Age doesn't stop you having a dream.

Mine is still to go into space. It was a mad idea when I was a child, and it is still a mad idea now, but I am not giving up on it.

After all, you've got to have a dream, right? What is life without one?

Epilogue

I have never written one of these before, so please forgive me as I explore new territory.

First, if you've made it this far – thank you. Thank you for joining me on this journey through time, space and memory. At the beginning of this book, I wanted to thank everyone I have interacted with in this universe – every friend, colleague, mentor and stranger who helped shape this version of my life. Now, I would like to include *you*, the reader, for spending a little of your own precious time here with me.

Back at the start, I said I was happy with my lot in life – and after writing about the voyage so far, I still stand by that conclusion. But, as life so often reminds us, the universe has its own way of testing our certainties.

While writing these pages – a process both enlightening and challenging – I was invited by my GP, Dr Denton, to have some routine blood tests. Most results were unremarkable, but a few were slightly off. We spoke on the phone, and I mentioned a period of unexplained weight

EPILOGUE

loss. I could rationalise some of it with what was happening in my life, but the timing felt suspicious. Dr Denton suggested a faecal immunochemical test – a simple at-home test that checks for trace amounts of blood in stool, a possible sign of bowel cancer.

The test came back positive. On its own that result isn't necessarily alarming, but combined with the weight loss it raised a genuine concern. I was bereft. As a family, we were already juggling a lot – new houses, changing circumstances, stress. Suddenly I found myself wondering how to explain to Lori that I might not be there for her future.

We talk about the future all the time. She's nearly sixteen now, and we imagine the milestones ahead: teaching her to drive, her first taste of adult freedoms, our birthdays are in March – when she turns eighteen, I'll turn 60. We have a mad, wonderful plan to take a gap year together, travelling the world and filming our adventures from two perspectives: *18 and 60* – or simply *1860*. I felt so honoured to be preparing for the Royal Institution's 200th anniversary Christmas Lectures, yet suddenly I was asking myself, *Would I be well enough to do them?* The next chapters of my life – quite literally – were fading at the edges.

I know how my mind works: I tend to dwell on things, to problem-solve my way out of uncertainty. Through my company I'd set up private healthcare, so I channelled the anxiety into action – seeking tests, pushing for appointments, trying to reclaim control through knowledge. Fear

EPILOGUE

of the unknown is natural, but I can face almost anything once I understand the challenge.

While filming *The Sky at Night* at ESA's European Astronaut Centre in Germany, I made countless calls between shoots, pretending I was chasing quotes for a new house. The team are wonderful, but until I had clear answers, I couldn't bring myself to share what was happening. It was one of the rare times I didn't tell Lori everything. I just needed to remove the uncertainty.

During filming, I was placed in a virtual-reality simulator – floating through a digital recreation of the International Space Station. It was exhilarating, disorientating, astonishingly real. Looking down at the curve of the Earth, I felt a pang: *Is this the closest I'll ever get to space?*

Meanwhile, the NHS was already in motion. With a positive FIT result, a colonoscopy is recommended within two weeks and, with my symptoms, a CT scan as well. Even while pursuing private care, I was impressed – and deeply grateful – that the NHS could move so swiftly.

After many calls, a wonderful woman named Katie, private secretary to my doctor Mr Dowson, arranged appointments for me the following week. The first test, the CT scan, went awry when the dye pooled under my skin instead of entering the vein, swelling my arm and halting the procedure. My one attempt to take control slipped away in that moment.

Next came the colonoscopy. You can choose sedation, but recovery takes longer, and I wanted to be home before

EPILOGUE

Lori finished school. Gas and air seemed the better option – and it appealed to my curious side. Why not take a *guided tour of my own intestines*? After all, A level biology was about to pay off.

Gas and air – a mix of nitrous oxide and oxygen – has a rather amusing effect on me. I've never taken drugs, but this must be the closest I've come to an out-of-body experience. Throughout the procedure, I laughed, giggled and peppered the staff with what I *hope* were semi-intelligent questions. The whole thing feels like a surreal blur now. Eventually, they asked me to stop puffing because I couldn't stop laughing.

Mr Dowson and the nurses were endlessly kind. Mr Dowson, with empathy that went above and beyond, understanding my fears, arranged a follow-up CT scan that same day and reviewed both sets of results immediately so I wouldn't have to wait.

The verdict: a clean bill of health. There were still minor questions about the blood results and weight loss, but overall, I was fine. *Roll on 1860.*

Yet the experience left me changed. It reminded me just how fragile life is, how quickly the story could have ended differently – how easily this could have been the last page in *this* version of reality.

I remain profoundly grateful for the version I inhabit: for the people I love, the work I do, the chance to still look up at the night sky and wonder. It sharpened my resolve to

EPILOGUE

live fully – to embrace the glorious chaos of existence and to make every day count.

As the Red Hot Chili Peppers so perfectly put it:

'This life is more than just a read-through.'

And I would add – so let's make the most of it, every single day, in this extraordinary universe we call home.

Acknowledgements

Writing *Starchild* has been one of the most illuminating and introspective journeys of my life. It has required me to revisit old memories, celebrate great joys and confront difficult moments with what I hope is honesty and on occasion courage. I could not have travelled this path alone.

First, my deepest thanks go to my daughter Lori, my compass and the brightest star in my sky. Your humour, kindness, wild imagination and unshakeable love carry me through every version of reality I inhabit. This book is as much yours as it is mine.

To my Mum, Sue, Hal and Gracie who have shared my journey from the beginning and to Martin, thank you for your love and support. To all my friends, especially Sue and Bas, Pip and Graham for your fun, laughter and the occasional nudge when the universe felt heavy. Your presence is woven through these pages.

Vicki, my long suffering agent, our journey continues but thank you for all of your help and support over the years.

My heartfelt thanks to the incredible teams I have worked

ACKNOWLEDGEMENTS

with – at the BBC, *The Sky at Night*, Imperial College, the Royal Institution, the University of Leicester and the many, many schools, festivals and communities who welcomed me so warmly. Your belief in the power of science communication fuels everything I do.

I am also grateful to the artists, photographers and institutions whose work appears in this book, especially Mattel who made me into a Barbie doll in 2023. This has been the ideal starting point for many discussions with children when talking about my career in science.

My sincere thanks to the editors and designers who helped shepherd *Starchild* into being. Albert for your patience and kindness and above all, coming up with the whole crazy idea of me writing my biography. Lucy, your guiding hand in the face of my chaotic thoughts, taking my words and shaping them into something readable, has been essential. Shammah and Marta, thank you for your guidance and forbearance. Thank you also to Issie and Morgana in the publicity team.

And finally, to every person who has ever attended a talk, asked a question, or shared their story with me – thank you. You remind me that curiosity is universal and that we are all, in our own way, trying to understand our place in the cosmos.

This book exists because of you – because of the people I've met in this version of reality, who have shaped my journey and made it brighter and more meaningful.